Quinta Nwanosike Warren

Energy and Sustainable Development

Also of Interest

Chemistry and Energy.
From Conventional to Renewable
Mark Anthony Benvenuto, 2026
ISBN 978-3-11-133090-7, e-ISBN 978-3-11-133109-6

Sustainable Products.
Life Cycle Assessment, Risk Management, Supply Chains, Ecodesign
Michael Has, 2024
ISBN 978-3-11-131482-2, e-ISBN 978-3-11-131546-1

Electrochemical Energy Storage.
Physics and Chemistry of Batteries
Reinhart Job, 2026
ISBN 978-3-11-914971-6, e-ISBN 978-3-11-161853-1

Power-to-Gas.
Renewable Hydrogen Economy for the Energy Transition
Méziane Boudellal, 2023
ISBN 978-3-11-078180-9, e-ISBN 978-3-11-078189-2

Sustainable Process Engineering
Gyorgy Szekely, 2024
ISBN 978-3-11-102815-6, e-ISBN 978-3-11-102816-3

Handbook of Electrical Power Systems.
Energy Technology and Management in Dialogue
Edited by Oliver D. Doleski, Monika Freunek, 2024
ISBN 978-3-11-126412-7, e-ISBN 978-3-11-126427-1

Quinta Nwanosike Warren

Energy and Sustainable Development

2nd revised edition

DE GRUYTER

Author
Dr. Quinta Nwanosike Warren
1629 K Street NW, Suite 300
Washington, DC 20006
USA
Quinta.Nwanosike.Warren@gmail.com

ISBN 978-3-11-164327-4
e-ISBN (PDF) 978-3-11-164348-9
e-ISBN (EPUB) 978-3-11-164343-4

Library of Congress Control Number: 2026930234

Bibliographic information published by the Deutsche Nationalbibliothek
The Deutsche Nationalbibliothek lists this publication in the Deutsche Nationalbibliografie; detailed bibliographic data are available on the Internet at http://dnb.dnb.de.

Cover image: pcess609/iStock/Getty Images Plus
Typesetting: Integra Software Services Pvt. Ltd.

www.degruyterbrill.com
Questions about General Product Safety Regulation:
productsafety@degruyterbrill.com

To the communities whose knowledge, creativity, and courage illuminate the path to sustainable energy. Your participation is not a footnote; it is the foundation.

If you want to go fast, go alone. If you want to go far, go together.

–Anonymous

Preface

The idea came from my time working on energy development projects with professionals from different disciplines. They were often very good in their areas of expertise whether it was finance, gender issues, economics, monitoring and evaluation, and so on. Very often, they lacked basic knowledge of energy, and they had to learn as the project progressed. As you can imagine, projects are much more successful if everyone involved understands enough basics to ask the right questions.

This book was written to help people, who are not energy professionals, understand the basics of energy. It covers energy topics from technical aspects of electricity generation and transmission to energy policy. It is written to inform but not overwhelm so that the material can be easily digested and the reader is armed with knowledge to speak intelligently about these topics.

The book ends with a framework for sustainable development, focused on energy but applicable to other sectors. A recurring theme is the necessity for stakeholder engagement in identifying both the problem and the solution. Finally, some examples of sustainable and unsustainable projects are presented.

Because most of my experience is with projects in Africa, the focus of the book is Africa. The general themes are applicable to other parts of the world but specific examples are always from Africa.

https://doi.org/10.1515/9783111643489-202

Contents

Acronyms

AC	Alternating current
BOO	Build-own-operate
BOOT	Build-own-operate-transfer
BOT	Build-operate-transfer
CCGT	Combined cycle gas turbine
CCS	Carbon capture and storage
CEO	Chief executive officer
CO_2	Carbon dioxide
COMELEC	Comité Maghrébin de l'Electricité or North African Power Pool
CSP	Concentrated solar power
DC	Direct current
EAPP	Eastern Africa Power Pool
EIA	Energy information agency
EOR	Enhanced oil recovery
FiT	Feed-in tariff
FMO	Netherlands Development Finance Company
GHG	Greenhouse gas
GWh	Gigawatt hour
HFO	Heavy fuel oil
IEA	International energy agency
IGCC	Integrated gasification combined cycle
IPP	Independent power producer
kWh	Kilowatt hour
LCOE	Levelized cost of electricity
LED	Light-emitting diode
LNG	Liquified natural gas
MCC	Millennium Challenge Corporation
MWh	Megawatt hour
NGCC	Natural gas combined cycle
NGO	Nongovernmental organization
NO_x	Nitrogen oxides
OCGT	Open cycle gas turbine
PEAC	Pool Energetique De L'Afrique Centrale or Central African Power Pool
PM	Particulate matter
PPP	Public–private partnership
PSP	Private-sector participation
PUE	Productive use of electricity
PV	Photo voltaic
REA	Rural electrification administration
SAIDI	System Average Interruption Duration Index
SAIFI	System Average Interruption Frequency Index
SAPP	Southern African Power Pool
SCADA	Supervisory control and data acquisition
SDG	Sustainability Development Goal

https://doi.org/10.1515/9783111643489-204

SO_x	Sulfur oxides
UN	United Nations
VOC	Volatile organic compound
WAPP	West Africa Power Pool
WTE	Waste-to-energy

Chapter 1
Introduction to Energy and Development

This book aims to equip development professionals – whose primary expertise may lie outside the energy sector – with a foundational understanding of key energy concepts. It explains, in accessible language, how technologies related to power generation, transmission, distribution, and carbon capture operate. Additionally, it introduces essential topics such as energy policy and the energy–water nexus. The goal is to enable development practitioners to engage more effectively with engineers and other energy sector experts, thereby contributing to the design and implementation of more effective and sustainable energy projects.

While the principles discussed are broadly applicable to development contexts worldwide, the book emphasizes African countries, using region-specific examples to illustrate key concepts.

This chapter introduces the critical role of energy in development and provides an overview of current energy use and electrification rates across Africa.

1.1 Energy Overview

The 7th United Nations' (UN) Sustainable Development Goals (SDGs) is "Affordable and Clean Energy" [1]. This SDG sets targets for increased access to reliable and affordable energy while increasing renewable energy. The UN estimates that in 2021, 9% of the world's population lacked access to electricity. This translates to roughly 675 million people without access to electricity in 2021. Energy use in developed countries is constant while it has almost doubled in developing countries over the 15 years leading up to 2019 [2]. Increasing energy use is to be expected in developing countries since development and population growth require increasing levels of energy use. In developed countries, most firms and individuals are connected already so energy use remains fairly constant. Additionally, in developed countries, the application of energy efficiency measures has contributed to the leveling of energy use despite increasing population.

Higher consumption of energy per capita has been shown to strongly correlate with higher GDP and increased life expectancy [3]. In 2022, electricity consumption in Africa was 0.6 MWh/capita, which is low when compared to 6.4 MWh/capita in Europe and 9.0 MWh/capita in North America in the same year [4].

The average annual population growth rate of sub-Saharan Africa in 2023 was the highest in the world at 2.5% compared to 0.8% in North America and 1.0% in South Asia [6]. Populations and population growth rates for different regions of the world are shown in Figure 1.1. The population of sub-Saharan Africa was third highest in the world in 2023 at 1.26 billion [5].

https://doi.org/10.1515/9783111643489-001

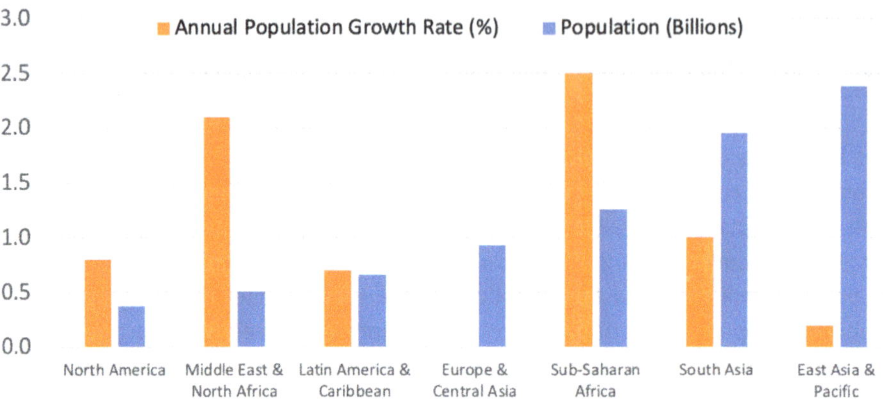

Figure 1.1: Population and population growth rates for different regions of the world in 2023. The growth rate for Europe and Central Asia in 2023 is zero [5].

Populations and population growth rates of individual countries in Africa are shown in Figure 1.2. The most populous country in Africa as of 2023 is Nigeria with 228 million people, followed by Ethiopia with 129 million people and Egypt with 114 million people.

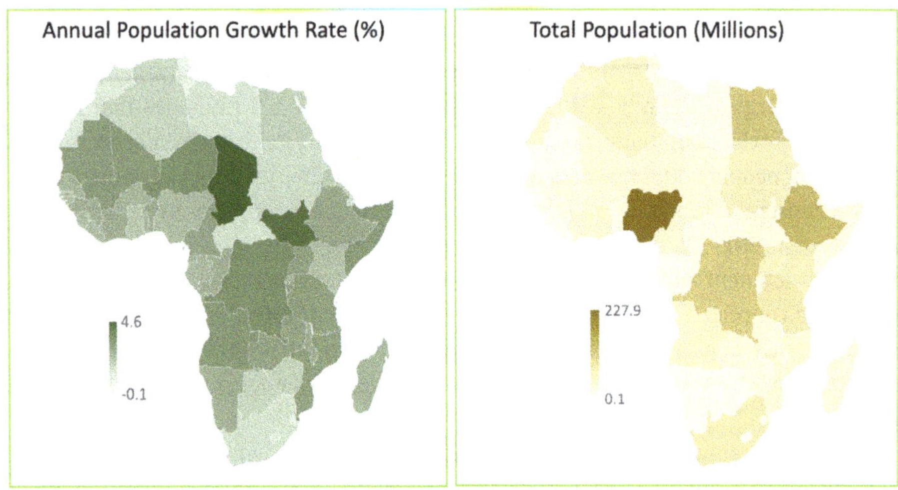

Figure 1.2: Annual population growth rates (%) and populations of African countries in 2023 [5, 6].

Approximately 573 million people in sub-Saharan Africa still lacked access to electricity when the SDG goals were set in 2015. While urban areas are better served, rural areas tend to have low access to the grid and with not enough economic activity to justify extension of the grid. The good news is that over time, access to electricity has increased noticeably despite increasing populations. This is illustrated for Africa in Figure 1.3.

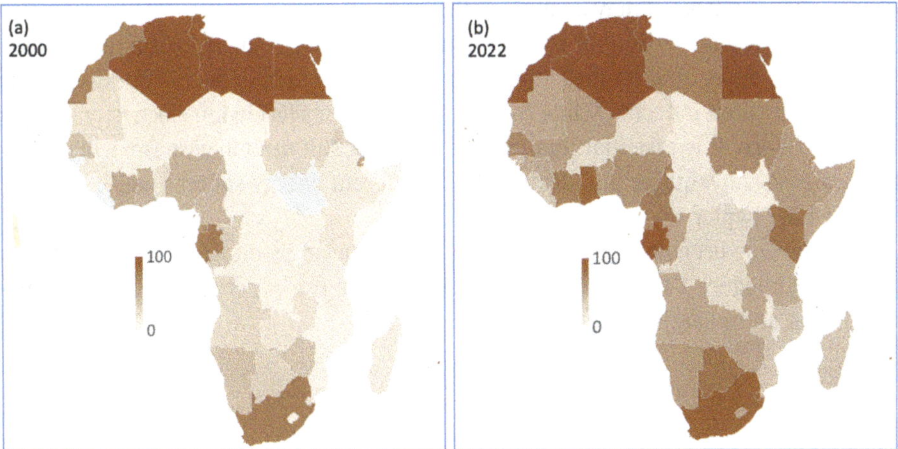

Figure 1.3: Percentage electrification rates in Africa in 2000 (Panel a) and in 2022 (Panel b). Gray indicates no data available [7].

The average access rate in Africa increased from 35% in 2000 to 58% in 2022. Eswatini, formerly called Swaziland, had the greatest increase in access rates which jumped from 20.4% to 82.3%.

Energy is an enabler for virtually every sector of the economy including health, agriculture, education, and transportation. Along with labor and capital, energy is an essential economic driver, powering productivity, innovation, and infrastructure. For instance, hospitals rely on electricity for life-saving equipment, clean water, and refrigeration of medication; farms use energy for irrigation, processing, and storage; and transportation systems depend on fuels or electricity to move people and goods. Without reliable access to energy, economic activity slows down and development is stifled.

In the household context, energy ensures comfort, health, and connectivity. Electricity enables lighting, cooking, heating, and the use of essential appliances like televisions, computers, and cell phones that facilitate communication, learning, and access to information. In the absence of electricity, families rely on traditional biomass such as wood, charcoal, and animal dung for cooking and heating. The use of these fuels leads to indoor air pollution, which is linked to respiratory illnesses like asthma and chronic bronchitis. Expanding access to clean and reliable energy not only improves health outcomes but also promotes economic resilience and social well-being.

There are three factors that are important when considering electricity delivery: availability, quality, and cost. Availability refers to whether or not there is a supply of electricity. This includes the presence of physical infrastructure needed to generate, transmit, and distribute electricity to end users. Quality refers to reliability, i.e., whether the electricity supply is constant or likely to be disrupted. Grid disruptions can add to the cost of doing business through equipment damage and disruption of production. Cost refers to the affordability of the electricity. If cost is too high, even if electricity is available, it may be inaccessible to some.

There is a wide variety of energy sources, each more suited for some purposes than others, depending on their characteristics. For example, petroleum dominates the transportation industry, while coal is most used for worldwide electricity generation. Therefore, it is essential that energy solutions suit the purpose for which they are designed. Advantages and drawbacks of different energy sources will be further explored in Chapter 3.

Worldwide, economies have historically been fueled by oil, coal, and natural gas as shown in Figure 1.4. In recent years, however, the use of renewable energy, especially wind and solar, is increasing.

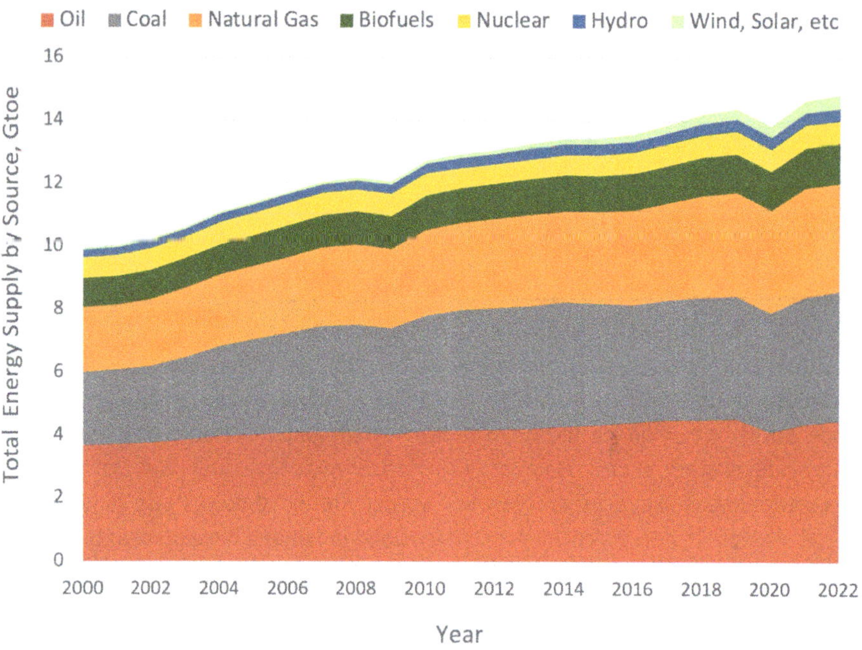

Figure 1.4: World total energy supply over time by source in gigatonnes of oil equivalent (Gtoe) [8].

1.2 Electrification Rates

Electrification rates reported for each country are national rates, i.e., a combination of urban and rural electrification rates. Figure 1.5 shows rural, urban, and national percent electrification rates in Africa in 2023. Urban electrification rates are usually much higher than rural electrification rates, which can be close to zero.

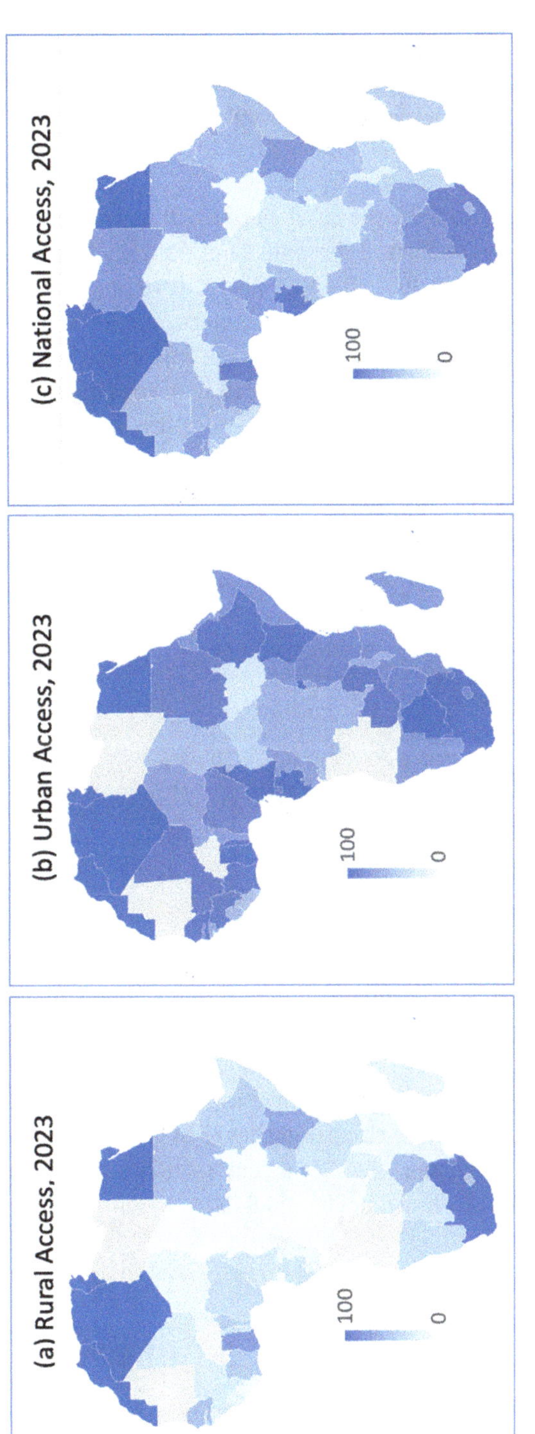

Figure 1.5: Chart showing how rural and urban percent electrification rates in 2023 are combined to get a national electrification rate. Gray indicates no data available [9].

Electrifying rural areas is difficult mostly because they usually have low population density and therefore are less profitable to electrify. Extending the electric grid is costly and the investment is not likely to pay off if there are not enough end users to buy the electricity it brings. Additionally, the rural population is less likely to be able to afford electricity.

Many solutions have been proposed to solve the rural electrification problem. In the United States, rural electrification was achieved through the establishment of the Rural Electrification Administration (REA) in 1936. The REA provided self-liquidating loans to local government and cooperatives for investment in electricity infrastructure projects. By the early 1970s, rural electrification rates in the United States had increased from 10% to about 98% [10]. This model can be adopted in developing countries but it requires the assumption that rural electrification is a good investment.

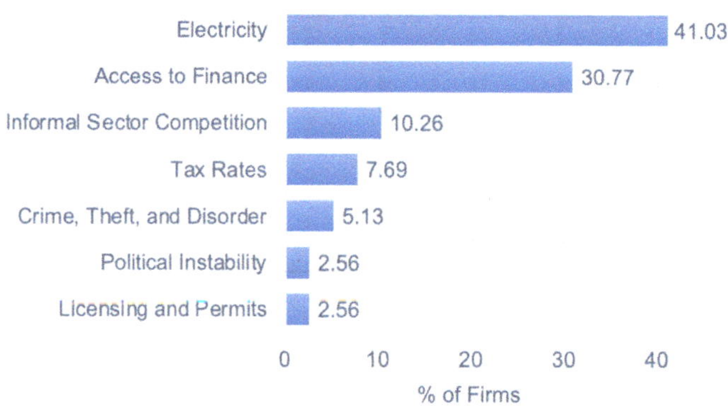

Figure 1.6: Operating constraints for firms in sub-Saharan Africa [11].

The development of microgrids in rural areas could replace extension of the national grid. These localized generation, transmission, and distribution systems are relatively cheap to set up and could still eventually be connected to the main grid.

Another potential solution is to encourage the development of enterprises in rural areas, which can create a more compelling case for investment by electric utilities and other investors. However, this presents a circular challenge as lack of electricity was identified by the World Bank as the most significant constraint faced by firms in sub-Saharan Africa from 2006 to 2010, as shown in Figure 1.6 [12].

Chapter 2
Sustainable Energy

This chapter covers the definition of sustainable energy and the three pillars of energy sustainability. It also covers factors that should be considered when developing sustainable development projects in the energy sector.

2.1 Pillars of Sustainable Energy

Energy sustainability is defined as addressing today's needs without compromising the ability to address future energy needs. Sustainable energy does not refer to renewable energy only. It includes the three pillars of economics, the environment, and the social aspect as illustrated in Figure 2.1. The needs of society have to be balanced against the well-being of society and the well-being of the environment. Renewable energy depends on fossil fuels and mining for materials. This interdependency precludes renewables from being the only sustainable energy source.

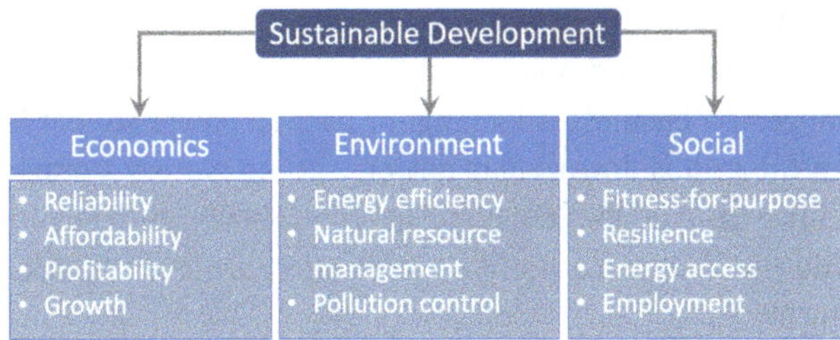

Figure 2.1: The three pillars of energy sustainability.

2.1.1 Economics

Economics generally boils down to the affordability of the energy solution. If it is priced too high, people will not be able to pay for it. If the electricity is being supplied by a utility, the utility also suffers as it will not have enough clients and its cash flow will be impacted. If the utility's cash flow is negatively affected, it will have fewer resources to maintain its electrical infrastructure, which would impact the quality of electricity delivery over time. Subsidies to make electricity more affordable fall under this pillar.

https://doi.org/10.1515/9783111643489-002

2.1.2 Environment

Taking care of the environment means reducing or eliminating pollution into the air, water, and soil. Polluted environments lead to illnesses in the community and loss of livelihoods for those who depend on the environment to make a living. For instance, fishermen cannot catch fish from polluted rivers, and farmers cannot grow crops on contaminated land.

Minimizing impacts on the environment also means using natural resources such as water for electricity generation efficiently. Efficiency refers to minimizing inputs while maximizing output and minimizing waste. For power generation, for instance, being efficient would mean using the least amount of fuel and water necessary while producing the maximum possible amount of electricity.

2.1.3 Social

The social pillar has to do with ensuring that the community has the job opportunities and a good standard of living. This pillar could include reliability, fitness-for-purpose, and resilience of the proposed energy solution. Reliability refers to the ability of the energy solution to deliver the required electricity when it is needed.

Fitness-for-purpose means that the energy solution should solve the problem it is supposed to address. For instance, if a business is looking for backup power so that it can run 24 h a day even when the grid is down, solar power with no battery storage would not be a good solution since the sun does not shine at night. Resilience refers to the ability of the energy solution to bounce back after a disruption like cyberattacks or natural disasters such as storms.

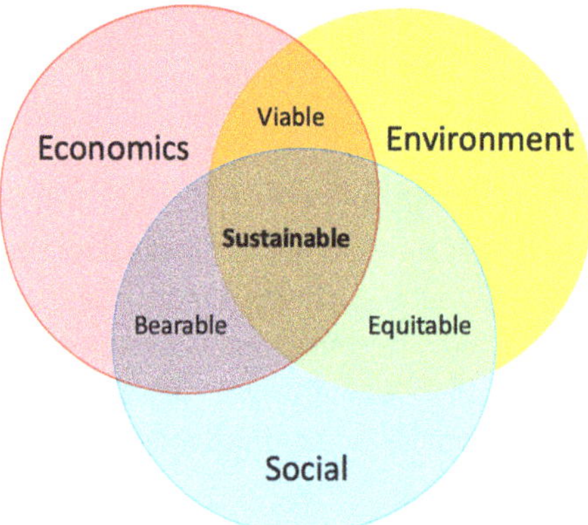

Figure 2.2: Interaction of the three pillars of sustainability.

The overlaps of the three aspects of sustainability are shown in Figure 2.2. The intersection of the environment and social is socio-environmental sustainability. It minimizes the impact of development and expansion of communities on the environment. This can be done through equitable policies which conserve the environment while supplying energy to society.

The intersection of economics and the environment is concerned with minimizing impact on the environment while providing affordable energy and making a profit and running a viable business.

Overlap of the economic and social deals with ensuring business practices has minimal negative impact on the community while providing opportunities for employment. Of course, where all three pillars overlap is where true sustainability lies.

2.1.3.1 Involve the Community

No one knows the needs of the community better than the community itself. While some solutions may have worked in other places, they should not be transferred to new places without being modified to fit the context of the new community. Solutions should not be crafted without the input of the communities they are meant to help. Failing to engage the community is a sure way to reduce the likelihood of the solution being effective and enduring.

2.1.3.2 Capacity Development

Most development work is done by international firms that are not based in the countries they do work in. Local firms are often excluded from participating in work being carried out in their own countries. Often this is because they lack the track record of working on large-scale projects or they have not been in business for long enough. One way to solve this problem and build their capacity would be to make it mandatory for the international firms to work with local firms similar to the way some government projects in the United States are set aside for small businesses. This would be a practical and sustainable way to develop human capacity, making it possible for local firms to gain experience necessary to bid on future development projects. Another positive outcome is that this approach would lead to higher education and training rates in the community, even for individuals not directly involved in international development projects.

2.1.3.3 Productive Use of Electricity

Productive use of electricity (PUE) is the act of engaging in income-generating activities through the use of electricity. Some development organizations will not fund projects if there is no guarantee that PUE will take place after the project is complete. It is unfair to expect that people getting electricity for the first time should not get it unless they use it to make money. Afterall, in developed countries, people are not held to that standard. Electricity is supposed to make lives easier and not add another burden to the backs of the people receiving it.

Energy is crucial for development but energy projects need to be implemented sustainably to have a lasting positive impact on the communities they are supposed to help. The three pillars of sustainability should be taken into consideration when developing energy projects.

Chapter 3
Energy Sources

This chapter covers the technical aspects of energy generation and transmission. The details of generation are covered for both renewable and nonrenewable energy sources. Processes are explained in as plain English as possible, and where technical terms are used, they are defined.

Energy sources can be divided into two types: nonrenewables and renewables. Renewable energy is energy that can be replenished within the lifetime of a human being. Examples include wind, solar, hydropower, geothermal, and biofuels. Nonrenewable energy forms include coal, natural gas, oil, and nuclear.

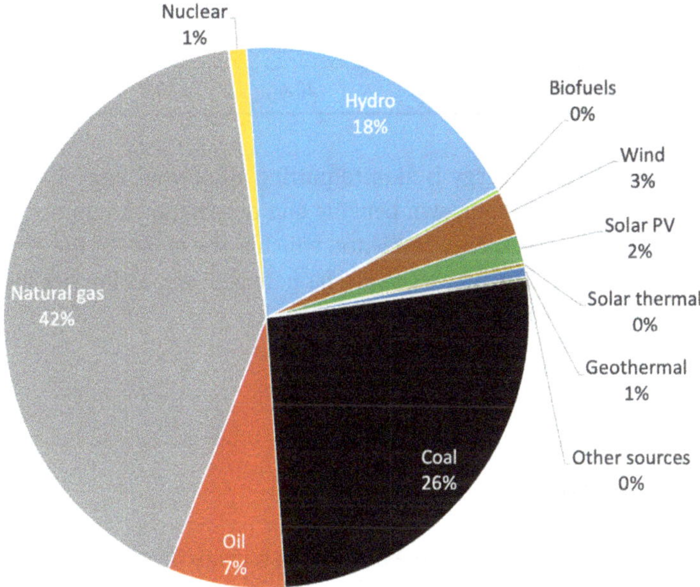

Figure 3.1: Sources of electricity in Africa in 2022 [13].

Across Africa, 42% of electricity generated came from natural gas, 26% from coal, and 18% from hydropower, according to the International Energy Agency (IEA) in 2017, as shown in Figure 3.1 and Table 3.1 [13]. The table also shows electricity losses during transmission through the electric grid. Solar technologies have been on the increase, facilitated by dropping costs for solar panels. Most developed economies were built using fossil energy such as coal and petroleum. Industrialization requires steady supplies of energy which renewables with their intermittent nature are unable to supply without battery backup.

https://doi.org/10.1515/9783111643489-003

Table 3.1: Electricity generation in GWh [13] and percentage transmission losses [14] from various energy sources across Africa from 2008 to 2022.

	2008	2011	2014	2017	2022
Coal	258,058	260,533	253,265	253,907	236,217
Oil	65,097	56,765	86,717	71,857	65,266
Natural gas	185,911	245,241	283,502	335,314	376,606
Biofuels	1,847	2,007	2,119	2,283	2,508
Nuclear	13,004	13,502	13,794	14,193	9,803
Hydro	100,038	113,081	126,518	129, 710	165,844
Geothermal	1,053	1,452	2,917	4,756	5,517
Solar PV	54	315	1,586	5,169	15,002
Solar thermal	0	0	0	1,101	2,308
Wind	1,295	2,432	5,548	11,564	24,510
Waste	2	6	0	0	0
Other sources	45	293	1,571	1,583	1,552
Total production (GWh)	**626,404**	**695,627**	**777,537**	**831,437**	**905,133**
Losses (%)	**22.6**	**21.3**	**20.4**	**19.4**	**19.3%**

Overdependence on one source of energy is akin to putting all of your eggs in one basket since all power sources have their own benefits and drawbacks. Therefore, a mix of energy types is important for energy resilience, which is the ability of the grid to bounce back from disruptions. Disruptions could be natural, such as hurricanes and wild fires, or man-made, such as cyberattacks.

3.1 Power Generation

3.1.1 How Electricity Is Generated

Power is usually generated by the turning of the blades of a turbine. The energy to do this can come from different fuel types such as wind, water, atomic fission, or the burning of coal, natural gas, biofuels, etc. Solar energy from photovoltaics (PVs) is one exception as there is no turbine involved. The turbine is attached to a generator which generates electricity when the blades of the turbine spin.

Some energy sources are more energy-dense than others. Table 3.2 shows the amount of energy in a kilogram of different energy sources. Uranium-235 is the densest energy source by far with over 22.7 million kWh/kg. Lithium-ion (Li-ion) batteries have the lowest energy density at 0.3 kWh/kg. More research and development are needed before batteries can compete with traditional fuel sources.

Table 3.2: Energy density of various fuel sources in kWh per kg [15, 16].

Energy source	Energy density (kWh/kg)
Li-ion battery	0.3
Coal	9.8
Diesel	12.5
Gasoline	13.5
Oil	14.7
Natural gas	15.3
Uranium-235	22,658,640

3.1.2 Nonrenewable Energy

Nonrenewable energy is energy derived from a source that cannot be replenished within the lifetime of a human being. These are typically fossil fuels but also include nuclear energy. Below are the most common nonrenewable energy sources.

3.1.2.1 Coal

Coal is compressed organic matter that is mainly carbon. There are four main types of coal namely lignite, sub-bituminous, bituminous, and anthracite. Lignite has the lowest energy content at 16.3 MJ/kg (4.5 kWh/kg), while anthracite has the highest energy content at 34.9 MJ/kg (9.7 kWh/kg). Coal types and their energy and carbon content are shown in Table 3.3. In Africa, the countries with the most coal power in 2022 were South Africa and Botswana as shown in Figure 3.2.

Another way of classifying coal is by categorizing it as either thermal coal or metallurgical coal. Thermal coal is primarily used in power plants to generate electricity. Metallurgical coal is used for steel production and is usually bituminous. It is generally higher quality, i.e., contains less sulfur, than thermal coal.

Burning of coal produces emissions such as particulates, mercury, sulfur dioxide, and different forms of nitrogen oxide. Sulfur oxides (SO_x) and nitrogen oxides (NO_x) give rise to acid rain which is damaging to the environment as well as buildings. The challenge of using coal to generate electricity is ensuring that these emissions are removed from flue gas, the gaseous byproduct of coal combustion, before it is vented in order to minimize impact on the environment.

Sulfur oxide compounds, or SO_x, are removed from flue gas using flue gas desulfurization units, which are also called scrubbers. Scrubbers use crushed limestone or lime to absorb sulfur compounds from flue gas. Mercury is also removed during this process. The product is called gypsum and it can be used for making cement and drywall.

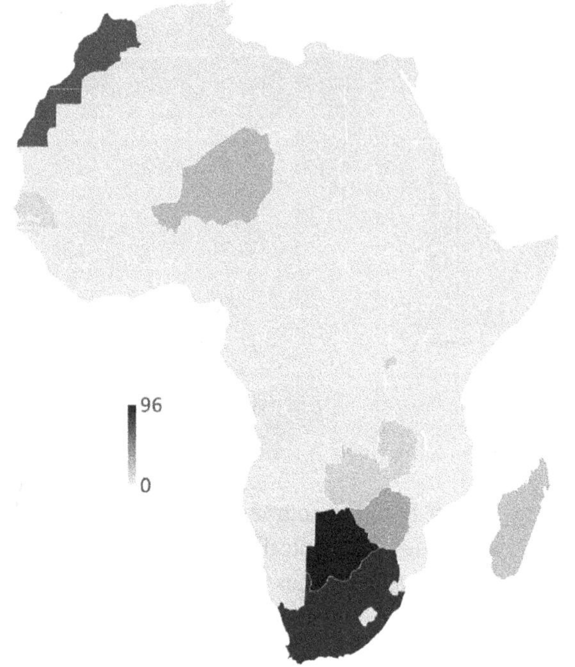

Figure 3.2: Percentage of electricity generated from coal in African countries in 2022 [17].

Table 3.3: Coal types and their energy and carbon content [18].

Coal type	Energy content (MJ/kg)	Carbon content (%)
Lignite	16.3	25–35%
Sub-bituminous	20.9	35–45%
Bituminous	27.9	45–86%
Anthracite	34.9	86–97%

Nitrogen oxides, or NO_x, cause smog and ground-level ozone which make the air hazy. Formation of NO_x can be prevented by reducing the air to fuel ratio in the combustion chamber. If NO_x is formed, it is reacted with ammonia in a catalyzed reaction and converted to nitrogen and water.

Particulates are noncombustible material left over from burning coal. Examples of particulates include PM_{10} and PM_{25}. PM_{10} is particulate matter that is less than or equal to 10 µm in diameter, and PM_{25} is particulate matter that is less than or equal to 25 µm in diameter. Smaller particulates are more harmful since they can more easily make their way around the body. Particulates can cause respiratory issues such as asthma and bronchitis if they are breathed in. They can be removed using an electrostatic precipitator. The particulates are given a negative charge. The positively charged plates of the precipitator attract the particulates. Particulates can also be removed using a fabric filter. Mercury is also removed in both processes.

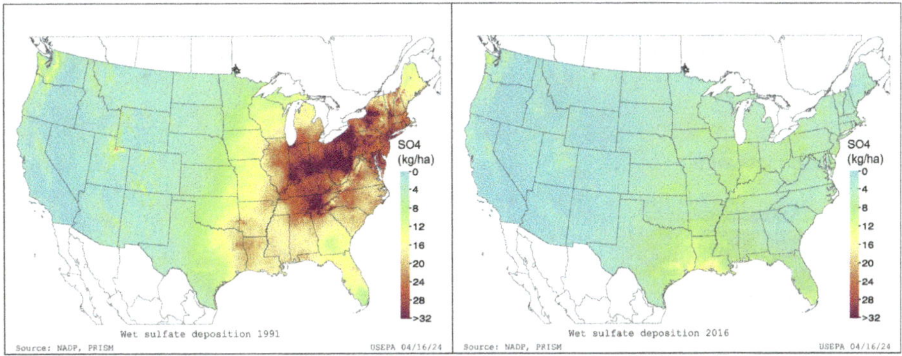

Figure 3.3: Acid rain in the United States in 1989–1991 and 2016–2018 [19].

Figure 3.3 shows the progress made in the United States in the fight against acid rain caused by sulfur compound emissions from coal burning for electricity generation. Technologies implemented in the 1990s led to a reduction in acid rain from 45 kg/ha/year to less than 15 kg/ha/year by 2010 in the eastern part of the United States.

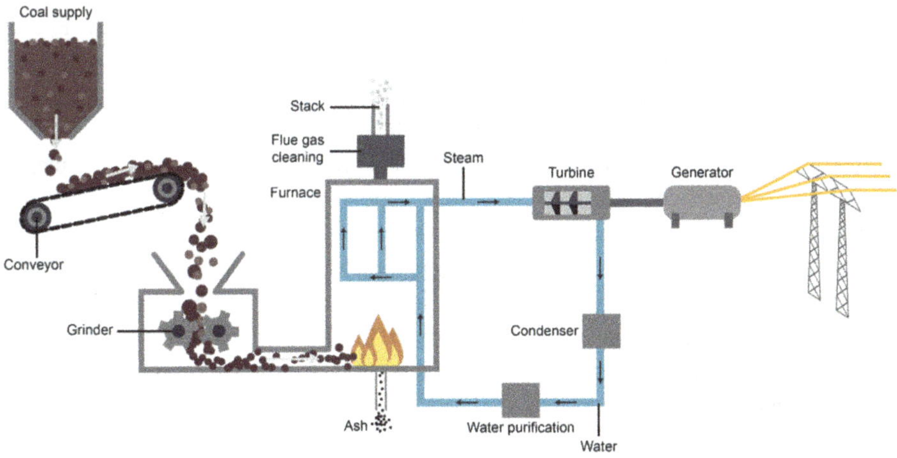

Figure 3.4: Schematic of electricity generation from coal.

The process by which electricity is generated from coal is illustrated in Figure 3.4. Coal is crushed to a fine powder to increase surface area and then blown into the combustion chamber of a boiler to be burnt in the presence of air. The heat generated is used to vaporize water in pipes lining the boiler. The high-pressure steam is passed over the blades of a turbine, causing the shaft of the turbine to turn at high speed.

The turbine shaft is connected to a generator which also turns rapidly, generating electricity. The steam condenses and is returned to the boiler. A schematic of a generator is shown in Figure 3.5.

Coal fly ash is the solid waste from a coal plant. It is generally stored in ash ponds which may be lined to prevent seeping of pollutants into the ground and subsequently into ground water.

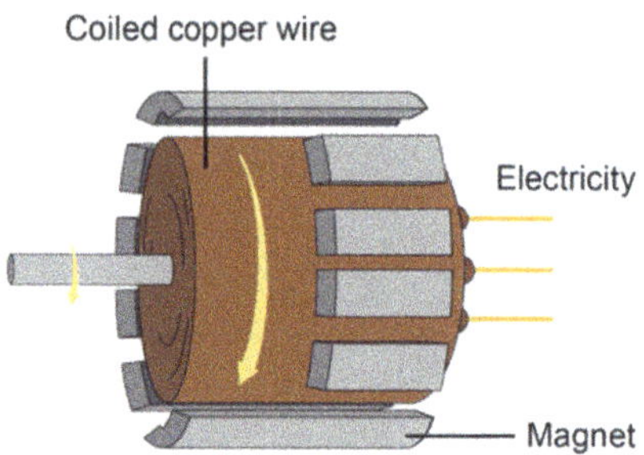

Figure 3.5: Schematic of a generator. Source: EIA [20].

The efficiency of a power plant is defined as the percentage of energy produced compared to the energy consumed from the fuel. The efficiency of a coal plant is about 30% due to thermodynamic limitations. The efficiency can be increased by operating at temperatures and pressures above the critical point of water which is 374 °C (705 °F) and 22 MPa (3,200 psi). Supercritical coal plants operate at 566 °C (1,050 °F) and 24 MPa (3,481 psi) and have an efficiency of about 42%. Ultra-supercritical coal plants operate at 760 °C (1,400 °F) and pressures up to 34 MPa (5,000 psi) and have an efficiency of about 44%. The increased efficiency means supercritical and ultra-supercritical coal plants produce more electricity but fewer emissions than conventional coal plants from the same amount of coal. The main drawback of more efficient plants is increased cost.

Integrated gasification combined cycle is a technology which uses gas derived from coal to generate electricity. The gas is a combination of carbon monoxide and hydrogen called synthesis gas or syngas for short. It can be burned to generate heat to convert steam to water to turn the blades of a turbine and generate electricity. Alternatively, syngas can be converted to methane which can also be combusted.

Coal plants use about 1,946 L of water per MWh of electricity produced [21]. The water is used to generate steam to turn the blades of the turbine. Water is also used to cool the steam leaving the turbine and returning to the combustion chamber. The resulting hot water is pumped to the cooling towers and allowed to cool. Some of it is lost to evaporation. The water is then reused for cooling. Heated cooling water may be disposed of rather than reused. A final use for water is to move coal fly ash to the ash pond. It should be noted that water use varies greatly depending on the specific configuration of the plant, e.g., whether water is recirculated or used once and disposed of, whether cooling is carried out using water or air, and whether the plant is subcritical or supercritical.

3.1.2.2 Natural Gas

Natural gas is mostly methane. In a simple gas turbine or open cycle gas turbine, natural gas is burned and the combustion gases expand, turning the blades of the turbine thus generating electricity. This process is illustrated in Figure 3.6. The efficiency of this process is between 20% and 35%. The efficiency can be increased by using the hot combustion gas to generate steam which is then used to turn another turbine. This process is called a combined cycle and can increase efficiency to 60%.

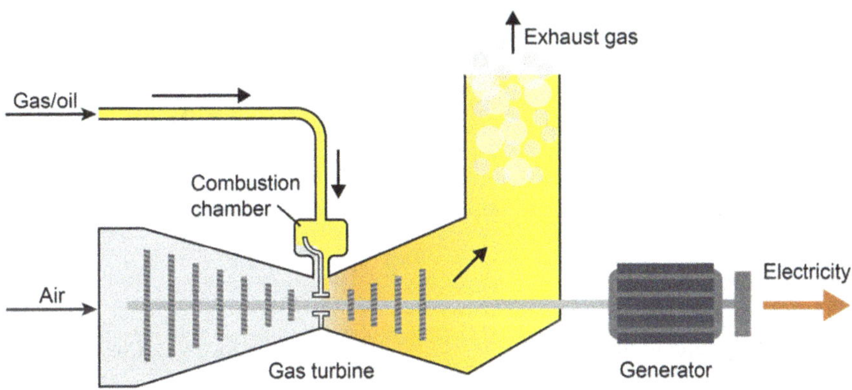

Figure 3.6: How a simple cycle, or open cycle, works.

Natural gas is much less polluting than coal but its use for power generation does produce NO_x. Due to its higher efficiency, the natural gas combined cycle (NGCC) produces much less CO_2 than a coal plant. Natural gas combined cycle plants use 492–1,136 L of water/MWh.

Due to sustained cheap natural gas prices, the number of natural gas plants has increased worldwide. In the United States, natural gas is displacing coal in the energy mix. In Africa, natural gas is accounted for roughly 42% of electricity generation in 2022 [13].

Liquified natural gas (LNG) is natural gas that has been cooled and compressed to a liquid for ease of transportation. LNG requires a regasification facility to turn the LNG back to gas for the purpose of electricity generation. Thus, an LNG plant must be of a few hundred MW capacity to be economical.

3.1.2.3 Petroleum

Petroleum can refer to heavy fuel oil (HFO) or diesel. HFO, also called residual fuel oil or bunker fuel, is relatively inexpensive as it is the very viscous waste product of

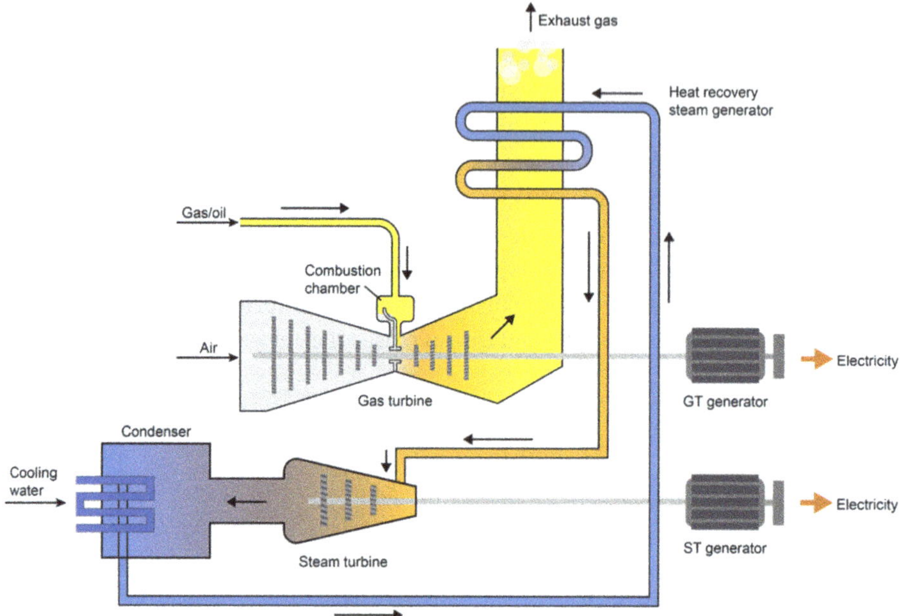

Figure 3.7: How a combined cycle works.

crude oil refining. Due to its high viscosity, it must be heated before use. Diesel is lighter and easier to handle but much more expensive.

HFO or diesel is combusted and the heat is used to turn water into steam. The steam is then run over the blades of a turbine to generate electricity, as demonstrated in Figure 3.7. Alternatively, exhaust gas from burning HFO can be used to drive the blades of a turbine.

The efficiency of electricity generation using HFO is 40–45%. If waste heat is used to heat up steam to generate additional electricity, efficiency can be increased to 50–55%.

Burning HFO and diesel can lead to emissions of SO_2, NO_x, mercury, volatile organic compounds, and particulates. To lessen the environmental impact, lower sulfur fuel oils can be used rather than HFO.

The use of HFO for electricity generation in the United States has decreased due to emission regulations and competing low prices from natural gas. In Africa, oil was used for generation of 7% of electricity in 2022. The percentage of oil use for electricity generation across Africa in 2022 is shown in Figure 3.8.

3.1.2.4 Nuclear

In a nuclear power plant, heat is generated from splitting of uranium atoms in a process called fission. Fission happens in the reactor core. The heat generated is regulated by control rods which absorb neutrons released during fission. The heat is used

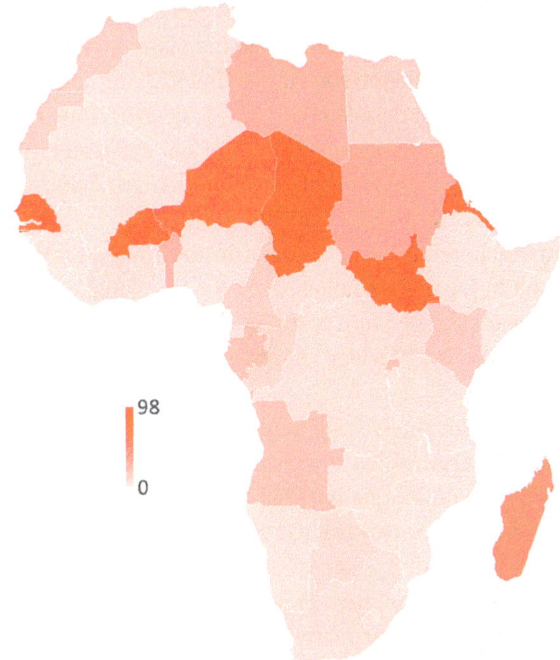

Figure 3.8: Percentage of electricity generation in African countries that came from oil in 2022 [22].

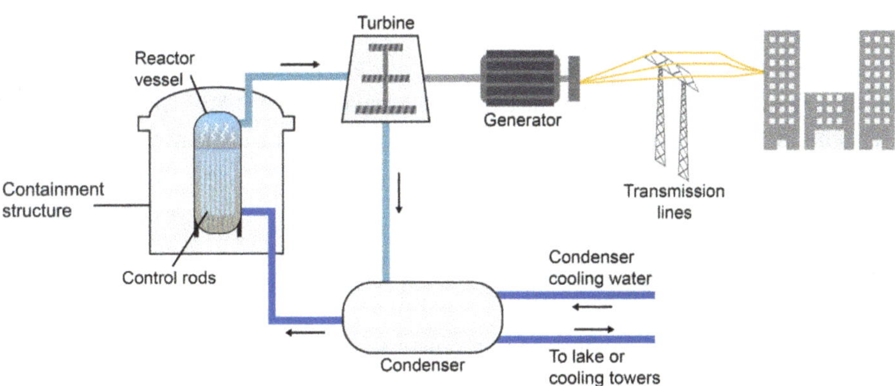

Figure 3.9: How a nuclear plant generates electricity.

to convert water to steam which turns the blades of a turbine thus generating electricity from the generator attached to the turbine. This process is shown in Figure 3.9.

Nuclear plants do not produce any carbon emissions although uranium mining and plant construction and decommissioning do release greenhouse gases. However, nuclear waste remains radioactive for hundreds of thousands of years. To prevent radiation poisoning, nuclear waste is buried at tens of meters or up to 5,000 m below the surface. The cost of waste disposal is often put on the public rather than the plant owner, at least in the United States. At Hanford Site in Washington state in the United States, 200,000 m^3 of nuclear waste is buried in 177 single- and double-shell tanks [23]. Over time, radiation has eroded the hulls and radioactive waste has seeped into the soil and water. Aquifers near the site are estimated to contain 1.0×10^9 m^3 of contaminated groundwater from the leaks. The site also contains 710,000 m^3 of solid radioactive waste.

Notable nuclear incidents include Three Mile Island in the United States, Chernobyl in Russia, and Fukushima in Japan. The fallout from the earthquake at Fukushima in 2011 has lasted over 13 years and cleanup is still in progress. The total cleanup cost is estimated to be $202.5 billion [24]. Nuclear energy is an expensive form of energy when accidents happen.

Nuclear plants take about a decade on average to plan and build. The capital cost is high, and specialized personnel are needed to run the plant. The Vogtle nuclear plant being expanded in the US state of Georgia for $36.8 billion after cost overruns is an example. Permits were filed in 2006 for two nuclear reactors of 1,250 MW each, but construction did not begin until 2013. Unit 3 began production in July 2023 and Unit 4 began production in April 2024. Construction took 15 years to complete, which is double the original estimate [25]. Because nuclear materials are subject to tight controls, there are certifications and regulations that must be followed during the planning and building as well as operation of the plant. This adds to the cost of nuclear energy. A nuclear plant could have 700 employees while a natural gas plant with the same output would only have a few dozen employees, adding to the operating costs.

Currently, South Africa is the only country in Africa with nuclear power. The country has two nuclear reactors at Koeberg nuclear power station outside of Cape Town. The first reactor was built in 1984 with a capacity of 930 MW while the second was built in 1985 with a capacity of 900 MW [26]. In 2022, the nuclear generators produced 4% of South Africa's electricity [27].

3.1.3 Renewable Energy

Renewable energy is energy that can be replenished within the lifetime of a human being. It made up roughly 24% of electricity generation in Africa in 2021 [13].

3.1.3.1 Biomass

Biomass is plant or animal matter. Biomass feedstocks include food waste, landfill gas, wood waste, animal manure, and agricultural residue. Biomass fuels can take many forms including biodiesel and biogas. Growing of plants to produce biomass may compete with food for farmland.

Generating electricity from biomass can be done in two ways. The first is to burn biomass and then use the heat to convert water to steam which then turns the blades of a turbine. This process has an efficiency of about 25%. Biomass can be burned and the hot combustion gases are used to turn the blades of a turbine.

The second way is to convert biomass to syngas, which is hydrogen and carbon monoxide, through a process called gasification. Gasification is the partial combustion of biomass with low levels of oxygen above 750 °C (1,382 °F). Syngas can be combusted to produce heat, or it can be further converted to methane which can also be combusted to produce heat. A subset of gasification carried out at lower temperatures around 500 °C (932 °F) is called pyrolysis. In this process, syngas is produced as well as bio-oil. The bio-oil can be combusted to produce electricity.

Biochar is already regularly produced and used in many parts of Africa. Biochar is produced by heating plant material in a low oxygen environment. It is generally used for cooking but may also be converted to biogas. The main disadvantage of biochar use for cooking is that it produces carbon dioxide, carbon monoxide, and other pollutants when burnt, causing respiratory diseases. High demand for biochar could lead to deforestation as trees and shrubs grow relatively slowly.

3.1.3.2 Hydroelectric Power

Africa had installed hydropower capacity of about 165,844 GWh in 2022. The countries with the greatest hydropower capacity are the Democratic Republic of Congo, Ethiopia, Zambia, Uganda, and Mozambique. The percentage of electricity generation in Africa from hydropower is shown in Figure 3.10.

Hydropower uses moving water to turn the blades of a turbine connected to a generator, thus generating electricity. This process is illustrated in Figure 3.11. The system may be run-of-the-river where the river current drives the turbine or a storage system where water is stored in a reservoir created by a dam on a river and released as needed to drive the turbines.

Pumped storage facilities may be used to store energy by pumping water to a higher elevation reservoir using cheap electricity and releasing the water to generate hydroelectricity during peak demand periods.

The efficiency of a hydropower plant is around 90%. Droughts can affect the amount of water available thus reducing the amount of electricity a hydroelectric power plant can generate. Floods bring silt and debris to the hydropower plant which can reduce storage capacity, compromise equipment, and clog intake gates, as happened with Lake Victoria flooding in Uganda in 2020 [29]. Excessive water can be removed from the reservoir by opening the gates and releasing water downstream. If dams are not well-maintained, they can burst and flood the surrounding areas.

While generation of hydroelectricity does not lead to the emission of pollutants, it can affect the environment in other ways. Dams can alter water temperature, chemistry, flow characteristics, and fish migration. The creation of a dam may result in the

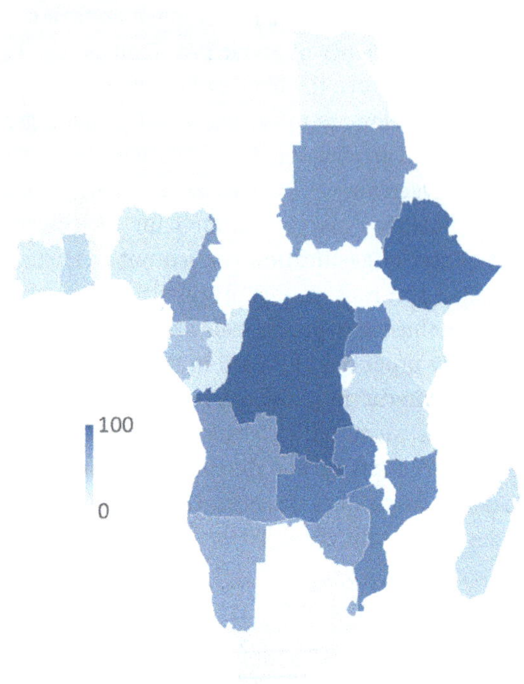

Figure 3.10: Percentage of electricity generation in Africa that comes from hydropower [28].

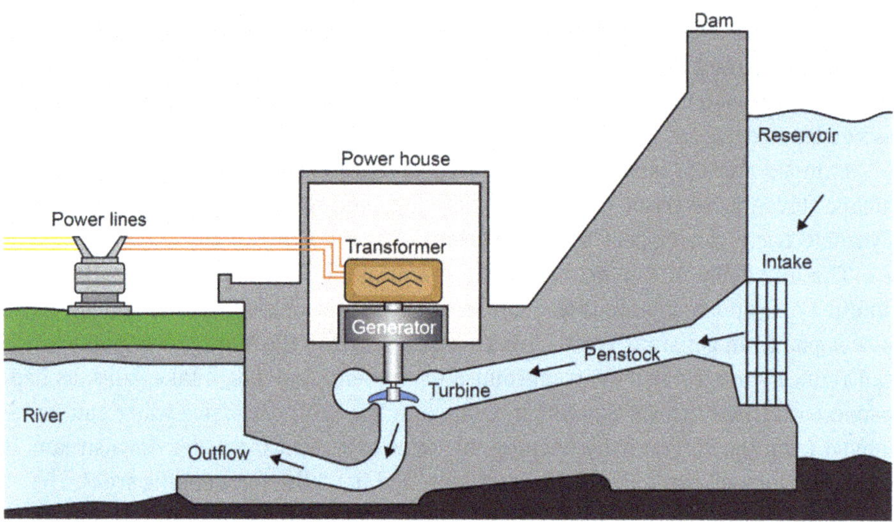

Figure 3.11: How hydroelectricity is generated.

relocation of people from the area. The reservoir may also flood previously dry areas thus altering the ecology of the region.

3.1.3.3 Geothermal

Geothermal plants harness heat energy from the earth and convert it to electricity. They require steam or hot water which is accessed from underground by drilling wells.

There are three types of geothermal power plants and they are illustrated in Figure 3.12. In a dry steam plant, naturally occurring steam from underground is used to turn turbines to generate electricity. When the steam cools and condenses, it is returned to the underground reservoir. In a flash steam plant, hot water at high pressure from underground is converted to steam which then turns the blades of the turbine. The cooled and condensed steam is then returned underground. In a binary cycle power plant, heat from the underground water at lower temperatures of 107–182 °C (225–360 °F) is transferred to a working liquid which is vaporized and used to turn the blades of the turbine. This working liquid is usually an organic liquid with a low boiling point. The cooled water is returned to the reservoir to be reheated. The efficiency of geothermal plants is about 10%. This can be seen in Figure 3.13 which shows the energy efficiencies of different types of power plants.

While geothermal energy is renewable, if the reservoir is not maintained, or if the rate of steam removal exceeds water return, the steam reservoir can become depleted. Thus, measures must be taken to ensure longevity of the reservoir.

Geothermal plants can produce a steady supply of electricity. They are cheaper to build and operate than coal or nuclear plants.

Not every part of the world has geothermal resources. In Africa, the greatest geothermal potential lies in the Great Rift Valley in countries like Kenya and Ethiopia [32]. In 2025, about 44%, or 754 MW, of electricity generated in Kenya was from geothermal sources [33].

Although geothermal energy is renewable, it has a few disadvantages. It has been linked to the occurrence of earthquakes caused by faults being lubricated by water being returned underground. One notable example was in November of 2017 when a magnitude 5.4 earthquake struck the city of Pohang in South Korea [34].

Some geothermal fields contain hydrogen sulfide, which is poisonous and corrosive, and sulfur dioxide, which can cause acid rain and respiratory diseases. Removing too much water for geothermal electricity production can cause subsidence of the ground due to reduced pressure in the reservoir.

3.1.3.4 Solar

Solar energy refers to harnessing the sun's energy and converting it to electricity. This is done most commonly through PV panels. A second way is through concentrated solar power (CSP).

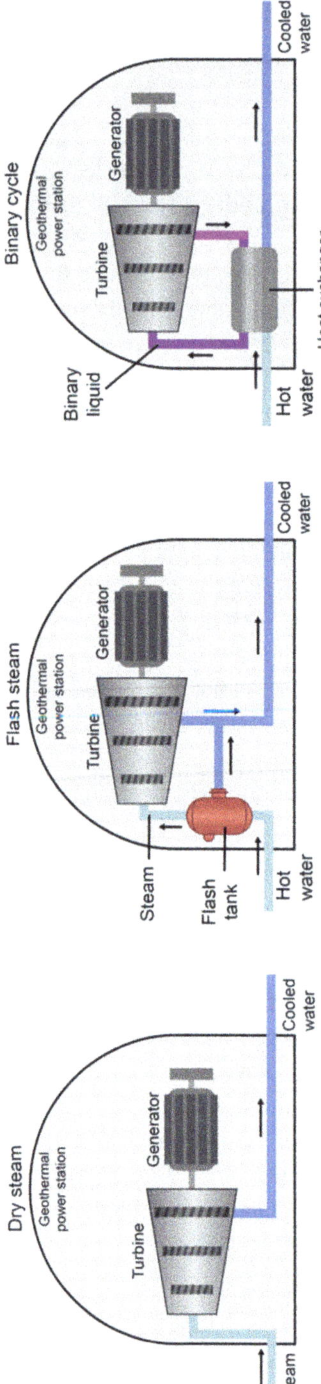

Figure 3.12: The three main types of geothermal plants.

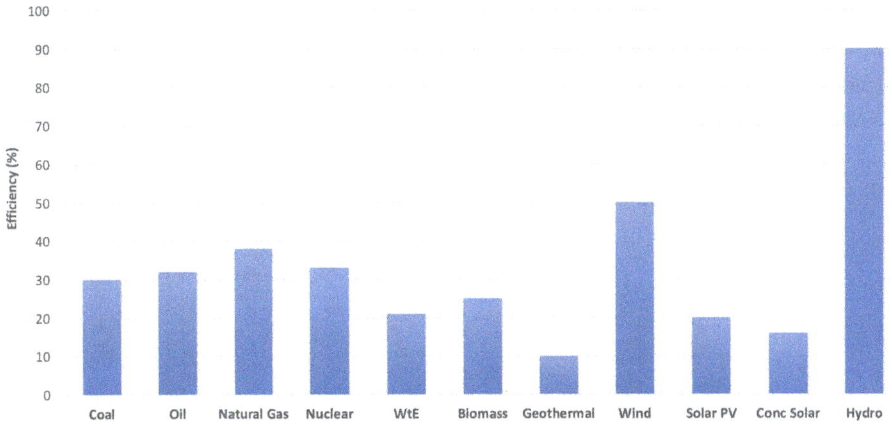

Figure 3.13: Efficiency of different electrical plants. WTE is waste-to-energy [30, 31].

3.1.3.4.1 Solar PV

PV cells are typically made of silicon which absorbs photons from sunlight, causing electrons in the cells to move, thus generating electricity in the form of direct current (DC). A solar inverter is then used to convert DC to alternating current (AC) which is needed to power appliances in the home. This mechanism is shown in Figure 3.14.

A solar panel has a maximum theoretical limit of 33.7%. In practice, the efficiency of solar PV is about 20%. Dust and dirt can reduce the efficiency of the panels. Efficiency drops about 0.5% every year after installation. Solar panels last about 25–30 years.

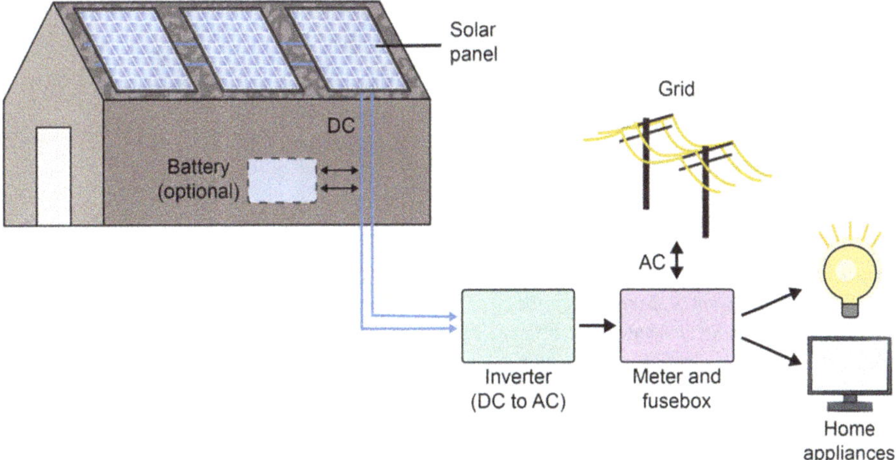

Figure 3.14: Typical solar system connected to the grid.

PV panels may be mounted on the roofs of buildings or on the ground. Solar panels take up a lot of space compared to other electricity generation methods. Roughly 9.3 m^2 are needed for every 1 kW of electricity. Direct solar irradiation across Africa per year is shown in Figure 3.15.

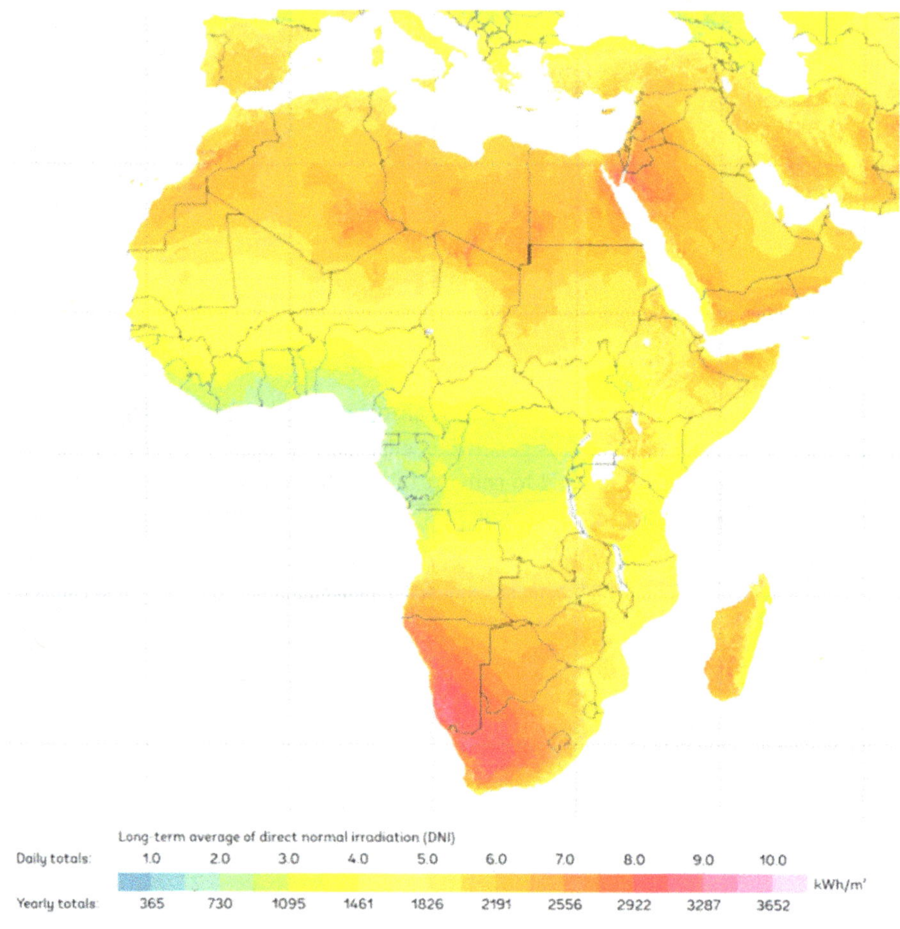

Figure 3.15: Direct normal solar irradiation in kWh/m^2 per year across Africa [35].
Source: *Data/information/map obtained from the "Global Solar Atlas 2.0," a free, web-based application developed and operated by the company Solargis s.r.o. on behalf of the World Bank Group, utilizing Solargis data, with funding provided by the Energy Sector Management Assistance Program (ESMAP). For additional information:* https://globalsolaratlas.info

The cost of PV has been declining in recent years, making solar more and more affordable. Costs for residential PV have fallen from $7.6/W in 2011 to $2.70/W in 2023 [36]. Costs over this time frame are shown in Figure 3.16.

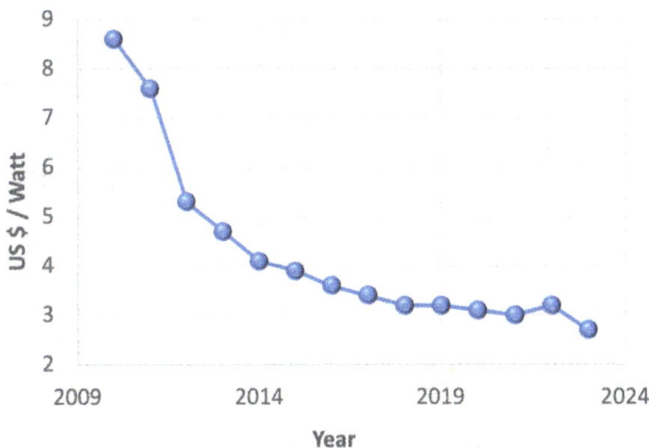

Figure 3.16: Cost of residential solar over time in US$ per watt.

3.1.3.4.2 Battery Storage

One disadvantage of solar energy is that it only works during the day when the sun is available. In places where regulations allow, solar power can be sold back into the grid. This practice is called net metering. It can be used as a form of storage. Excess power is sold into the grid during the day, and power can be bought from the grid at night.

Batteries are another form of storage. They can be charged during the day with excess electricity when solar energy is available and can be discharged at night to supply electricity. Batteries make it possible for complete disconnection from the grid or setting up of mini-grids. The most common batteries are Li-ion batteries.

The capacity of a battery is expressed in megawatts and the time it takes for the battery to discharge completely. For instance, a 100 MW battery with 5 h of storage can supply 100 MW over 5 h or 50 MW over 10 h. The longer the storage capacity in terms of discharge time, the lower the cost per MWh.

The specific energy of Li-ion batteries is about 0.36–0.875 MJ/kg. They leak energy at a rate of 0.35–2.5% per month, so may discharge completely if left unused for too long. Temperature, current of charging or discharging, and depth of discharge all affect the cycle of the Li-ion battery. If damaged or charged incorrectly, the Li-ion battery can explode or catch fire. While the use of batteries does not directly result in emissions, mining for lithium, nickel, cobalt, and other metals does significantly impact the environment. Manufacturing of a 1-kg Li-ion battery uses about 67 MJ of energy. The process releases between 62 and 140 kg CO_2 equivalent per kWh.

Li-ion batteries typically last 10 years. The disposal of batteries is becoming a much talked about issue. Some battery companies allow clients to return their old batteries to them for recycle or disposal.

3.1.3.4.3 Concentrated Solar Power (CSP)

CSP or thermal solar is carried out using lenses or mirrors to concentrate energy from the sun to generate steam to turn the blades of a turbine and generate electricity. CSP with storage heats up a liquid, such as molten salt or synthetic oil, which can store heat for hours. As such, CSP with storage can generate electricity even at night. The efficiency of CSP systems ranges from 7% to 25%.

Trough systems utilize a parabolic mirror with a liquid-filled pipe running along the center. Focused sunlight heats the liquid. If the liquid is oil, it can be heated to as high as 400 °C (750 °F). The hot liquid is then used to heat water to steam for electricity generation.

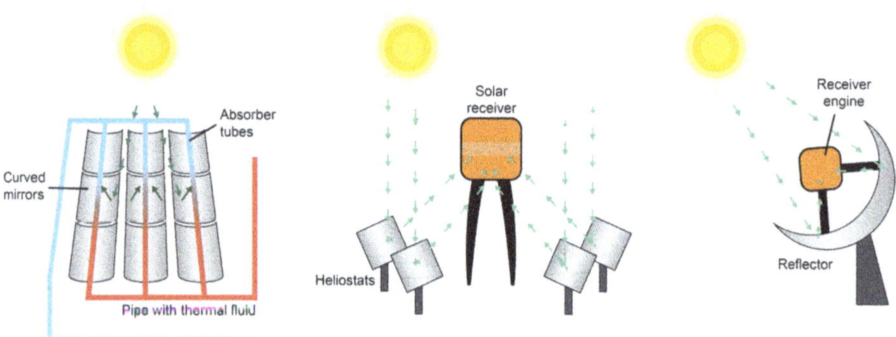

Figure 3.17: Different types of concentrated solar power (trough, power tower, and dish).

Power tower systems use flat mirrors to concentrate sunlight onto a receiver on a tower. The receiver heats up a liquid which then heats up water to steam for electricity generation. If molten salt is used, it can be heated to temperatures as high as 566 °C (1,050 °F). Molten salt can retain heat for many days so it is an efficient storage medium. It can be used to generate electricity when sunlight is not available.

In dish systems, a receiver is mounted at the focal point of a mirrored dish and connected to a combustion engine. The engine uses heated helium or hydrogen to drive a generator. The dish tracks the sun as it moves across the sky. The trough, power tower, and dish systems are illustrated in Figure 3.17.

The largest CSP plant in the world is the fourth phase of Mohammed bin Rashid Al Maktoum Solar Park in Dubai in the United Arab Emirates. This phase combines CSP and PV with a total CSP power production of 600 MW from a parabolic trough system and 100 MW from a solar tower [38]. The tower was completed in 2021, and the trough in 2023. Noor Power Station in Morocco is the world's second largest CSP power plant at 510 MW. The plant uses molten salt to store energy with a total storage capacity of 17 h. Noor I and II use a parabolic trough design, while Noor III uses a power tower system. The station was commissioned in 2016.

3.1.3.5 Wind

A wind turbine generates electricity by the wind turning the blades of the turbine, thus producing electricity through the attached generator.

Wind speeds around Africa are shown in Figure 3.18. They are strongest around the southern coast and over northern Africa.

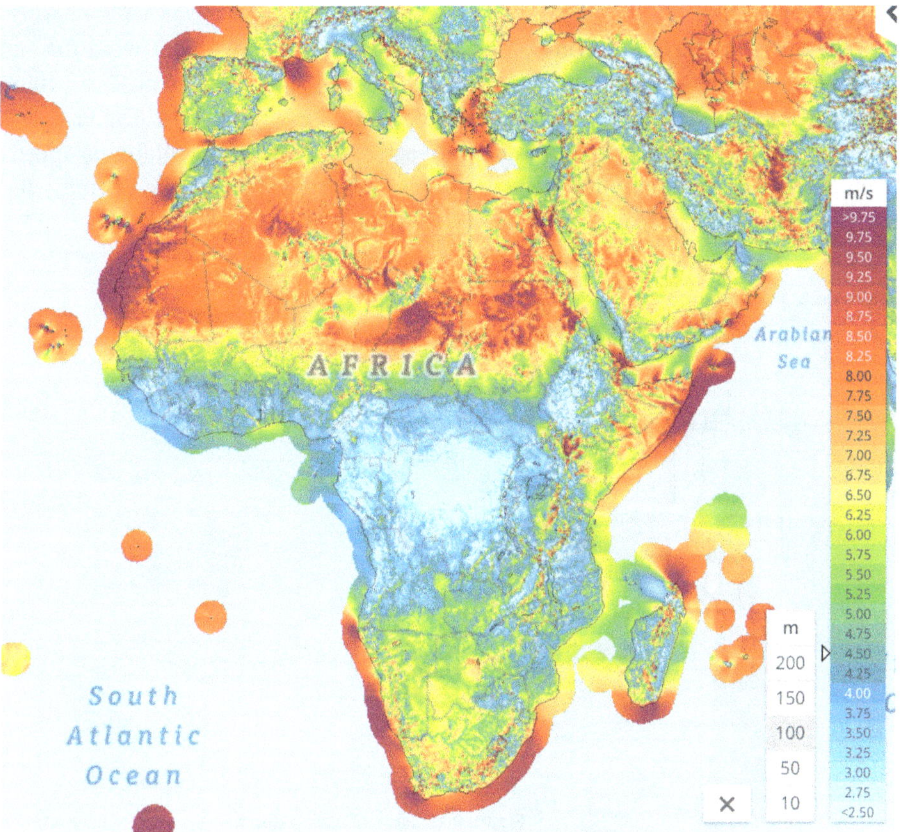

Figure 3.18: Wind speed around Africa in meters/second [39].
Source: *Data/information/map obtained from the "Global Wind Atlas 3.0," a free, web-based application developed, owned and operated by the Technical University of Denmark (DTU). The Global Wind Atlas 3.0 is released in partnership with the World Bank Group, utilizing data provided by Vortex, using funding provided by the Energy Sector Management Assistance Program (ESMAP). For additional information:* https://globalwindatlas.info.

Wind power is directly proportional to the cube of wind speed. It is also proportional to the swept area of the turbine and the density of air:

$$P = \rho A v^3$$

where P is the power, ρ the density of air, A the swept area, and v the wind speed.

This means that the lower the altitude, and the bigger the blades of the turbine, the greater the power generated. The wind speed is the greatest factor since wind power is dependent on the cube of wind speed. The faster the wind, the more the power generated.

The different parts of a wind turbine are shown in Figure 3.19. A wind turbine sits on a steel tower about 80 m tall and the blades can span up to 79 m. The turbine is attached to a generator through a low-speed and high-speed shaft. The gear box connects the low-speed shaft to the high-speed shaft. It increases rotational speeds from 30 rotations per minute (rpm) to about 1,800 rpm. The nacelle contains the shafts, the gearbox, and the generator.

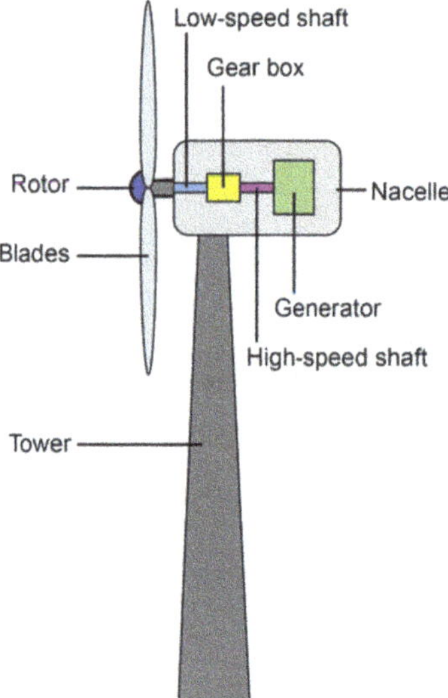

Figure 3.19: Parts of a wind turbine.

The turbine rotates to face the strongest wind. Wind turbines start to generate electricity when the wind speed is about 9.7 km/h (6 mile/h). If the wind blows too hard, say 88.5 km/h (55 mile/h), the turbines will shut down to prevent damage. The efficiency of wind power is about 50%. Wind turbines last for 20–25 years.

The capital cost for offshore, i.e., in water like the ocean, wind power is very high. Capital and operating costs for onshore (on land) wind are low compared to combined cycle gas plants.

Wind energy does not give off any emissions. The blades of a wind turbine are made of composite polyester reinforced with fiberglass or carbon fiber. Disposal of wind turbines at the end of their life is done by thermal recycling to recover the fiberglass, shredding them for filler during cement production or throwing them into a landfill.

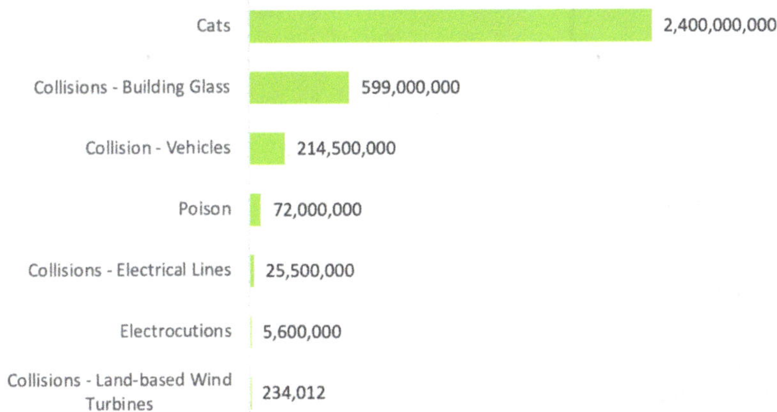

Figure 3.20: Annual bird mortality from select anthropogenic causes in the United States.

Wind turbines are said to be dangerous for birds. Data from the US Fish and Wildlife Service shows that cats kill around 2.4 billion birds every year and collisions with buildings kill 599 million birds annually. Wind turbines kill only about 234,000 birds every year as shown in Figure 3.20. However, wind turbines may disproportionately affect birds of a specific species due to migration [40]. Using maps of bird habitats and migration patterns during planning of wind turbines can help to mitigate killing of birds.

3.1.3.6 Waste-to-Energy

Waste-to-energy (WTE) is the conversion of waste to electricity, fuel, or heat. There are various ways in which this conversion can be done. These processes are the same as the processes used to convert biomass to electricity. The process of electricity generation from a typical waste-to-energy plant is shown in Figure 3.21.

Incineration is the combustion of waste organic material. The heat is used to convert water to steam, and the steam is used to turn the blades of a turbine and generate electricity. The efficiency of this process is about 14–28%. The process produces emissions such as particulates, SO_x, and fly ash.

Gasification is the conversion of organic waste in a low oxygen environment at temperatures greater than 700 °C (1,292 °F) to syngas (carbon monoxide and hydrogen) and carbon dioxide as a byproduct. Syngas can be combusted to generate heat which can be used to convert water to steam for electricity generation.

Pyrolysis is the decomposition of organic matter in a low oxygen environment at 300 °C (572 °F) to 700 °C (1,292 °F) to produce syngas, bio-oil, and char. It is the first step

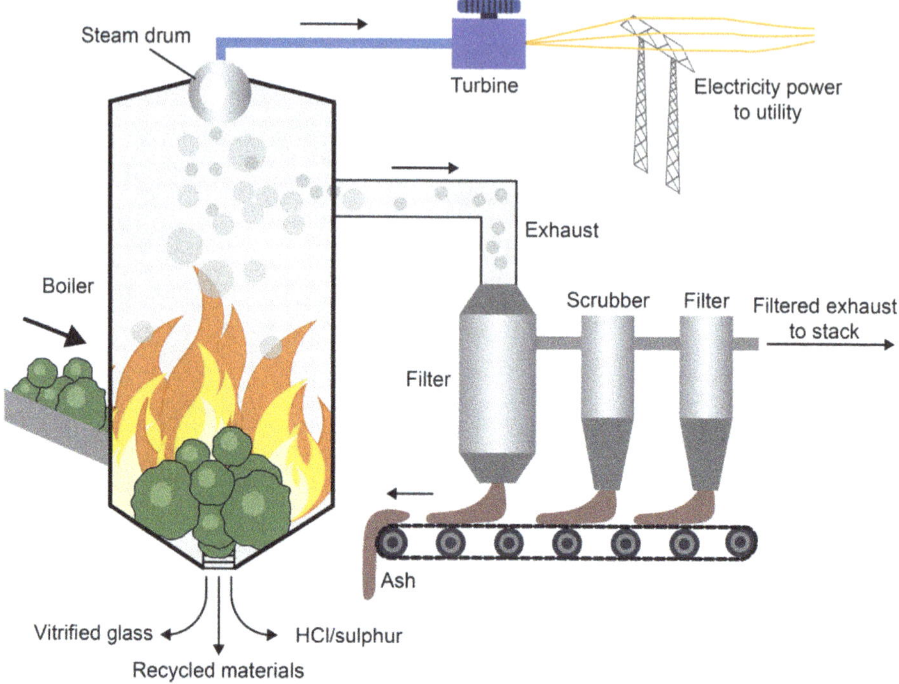

Figure 3.21: Electricity generation from a typical waste-to-energy power plant.

in the gasification process. Pyrolysis can be used to treat plastic waste and municipal solid waste. Syngas and bio-oil can be combusted to generate electricity. Char is used to make activated carbon for water filtration or to improve soil fertility for agriculture.

3.1.3.7 Other Energy Sources

Other methods of electricity generation exist that are still in development. They include wave energy which uses tides or water currents.

Hybrid power systems involve using more than one type of power generation. This is an excellent way to mitigate the disadvantages of individual generation types. For instance, a solar power system could be combined with a diesel generator. The solar system could provide power during the day, while the diesel generator could be used at night and on cloudy days.

The benefits and drawbacks of different forms of electricity described in this chapter are shown in Table 3.4.

Table 3.4: Benefits and drawbacks of various forms of electricity generation.

Power source	Benefits	Drawbacks
Coal	– Energy dense	– SO_2, NO_x, and mercury, particulates – CO_2 emissions – High water use
Natural gas	– Cheap – Fast to build – Dispatchable (power output can be quickly adjusted)	– Price volatility
Oil	– Cheap (heavy fuel oil)	– SO_2, NO_x, and mercury, particulates – CO_2 emissions
Biomass	– Wide variety of fuel sources	– Competes for farmland – Long grow times
Hydro	– Dispatchable – No emissions	– Damage to ecosystems – Resettlement of people – Danger if dam breaks – Water diverted from users downstream – Impacted by rainfall
Geothermal	– Cheap to build and operate – Low emissions	– Can cause earthquakes – Declining capacity over time – Low efficiency
Solar	– Low end-user tariffs – No direct emissions	– Intermittent – Requires sunlight – Sunlight amounts vary – Vast land area needed
Wind	– Relatively cheap (onshore)	– Intermittent – High capital cost for offshore – Variable wind speeds
Nuclear	– Steady electricity production	– Radioactive waste – High capital cost – Long build times – Highly skilled operators required
Waste-to-energy	– Reduces waste in landfills	– Harmful emissions

3.2 Transmission and Distribution

Electricity flows from the generation plant to transmission lines and then to distribution lines. Transmission lines transport electricity at high voltages over long distances from generation plants to substations close to electricity consumers. The voltage level of transmission lines is over 110 kV. Distribution lines transport electricity over short distances at voltages less than 40 kV. The voltages between 40 and 110 kV are considered sub-transmission voltages.

The generator at a power plant is connected to a step-up transformer which increases the voltage of the generated electricity. The voltage of electricity is increased, above 110,000 V and up to 765,000 V depending on the distance it will be transmitted. Increasing the voltage improves the efficiency of transmission, thereby reducing losses. Before electricity is distributed to homes, the voltage is reduced to a safer level of 100–250 V through a step-down transformer. The generation, transmission, and distribution of electricity are illustrated in Figure 3.22.

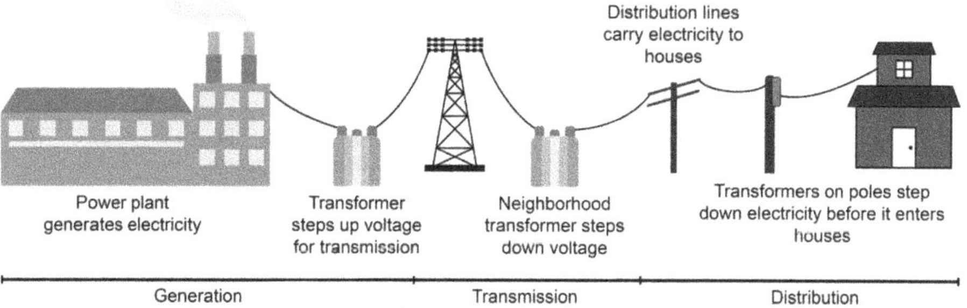

Figure 3.22: Generation, transmission, and distribution of electricity.

The grid must be perfectly balanced, i.e., demand, also called the load, must perfectly match generation. If there is an imbalance, the grid goes down and an outage occurs. Power plants must have reserve generating capacity, called spinning reserves, to buffer unexpected gaps between demand and supply.

The intentional shutdown of the grid is called load shedding or a rolling blackout and may be implemented due to insufficient electricity supply or due to lack of adequate infrastructure to move electricity to where it is needed. A brownout is a drop in grid voltage which may be planned or unplanned. Brownouts occur when demand is close to the maximum capacity of the grid. An intentional brownout prevents a blackout, which is a widescale interruption of power.

Transmission lines can be overhead or underground. Underground cables are much more expensive. They are also more difficult to service since they are buried. However, they are more protected from the elements and less susceptible to damage from wind.

The grid must be reliable so that electricity can be deployed to where it is needed when it is needed. It must also be flexible to allow for a diversity of sources to feed in electricity from wherever it is generated.

One measure of grid reliability is called System Average Interruption Duration Index (SAIDI). SAIDI is the duration of interruptions in electricity supply over a set period of time. Another measure of grid reliability that is often used is System Average Interruption Frequency Index (SAIFI). SAIFI measures how often a consumer experiences sustained outages over a set period of time. Values for these indices in the United States in 2017 were 7.8 h for SAIDI and 1.4 interruptions for SAIFI [41].

A distributed network is one where electricity is generated near where it is consumed. The electricity is deployed over a distribution network rather than the transmission network. Distributed generation technologies include solar PV on individual buildings, diesel generators, and small wind turbines. The advantage of using distributed systems is that electricity does not have to travel far so there are fewer losses. Microgrids are small-scale grids which can operate independently from the main grid, which helps mitigate issues on the main grid. Microgrids contain one or more types of distributed energy. Distributed systems and microgrids decrease the need for transmission lines. At the very least, they provide electricity while the main grid is still being designed and constructed.

Generation, transmission, and distribution of AC are typically done via three-phase or single-phase. In three-phase, power is supplied through three conductors whereas with single-phase, it is supplied through one conductor. Three-phase is typically used for industrial consumers such as factories and data centers. It has fewer losses and is more efficient, delivering power at a steady rate. Single-phase uses less power and is used for smaller loads. It is cheaper due to less complex designs. Single-phase systems can be derived from three-phase systems.

3.3 Levelized Cost of Electricity (LCOE)

Levelized cost of electricity (LCOE) is the average net present cost of electricity generation over the lifetime of a generation plant. It combines capital, operating and maintenance, performance, and fuel costs and provides a common basis for comparing the cost of electricity production for different plant types. However, LCOE does not account for how well a plant can match electricity demand such as in the case of renewable energy sources.

LCOE can be calculated as follows [42]:

$$\text{LCOE} = \frac{\text{Sum of costs over lifetime}}{\text{Sum of electrical energy produced over lifetime}} = \frac{\sum_{t=0}^{n} \frac{C_t}{(1+r)^t}}{\sum_{t=0}^{n} \frac{E_t}{(1+r)^t}}$$

where n is the expected lifetime of the system in years, t the year, C_t the cost in year t, r the discount rate, and E_t the electricity generated in year t.

The cost C_t includes capital costs, operating and maintenance costs, fuel costs, as well as any negative costs such as incentives and tax credits.

The lower the LCOE, the better. The capacity factor of a power plant is the ratio of actual electrical energy produced over a set period, say a year, to the maximum electrical energy capacity of the plant over that same period. A higher capacity factor increases the electrical energy produced over the lifetime of the plant, thus reducing the LCOE. The LCOE of various plant types is shown in Table 3.5 and Figure 3.23.

Table 3.5: Levelized cost of electricity (LCOE) for new generation sources in 2021 dollars per megawatt hour. Tax credits and other incentives are not included. LCOE values are based on a 30-year cost-recovery period. Source: EIA Annual Energy Outlook 2022 [43].

Plant type	Capacity factor (%)	Levelized capital cost ($)	Levelized fixed O&M[1] ($)	Levelized variable O&M ($)	Levelized transmission costs ($)	Total system LCOE ($/MWh)
Dispatchable technologies						
Ultra-supercritical coal	85	52.11	5.71	23.67	1.12	82.61
Combined cycle	87	9.36	1.68	27.77	1.14	39.94
Combustion turbine	10	53.78	8.37	45.83	9.89	117.86
Advanced nuclear	90	60.71	16.15	10.30	1.08	88.24
Geothermal	90	22.04	15.18	1.21	1.40	39.82
Biomass	83	40.80	18.10	30.07	1.19	90.17
Hydroelectric	54	46.58	11.48	4.13	2.08	64.27
Resource-constrained technologies						
Wind, onshore	41	29.90	7.70	0.00	2.63	40.23
Wind, offshore	44	103.77	30.17	0.00	2.57	136.51
Solar photovoltaic	28	26.60	6.38	0.00	3.52	36.49
Capacity resource technologies						
Combustion turbine	10	53.78	8.37	45.83	9.89	117.86
Battery storage	10	64.03	29.64	24.83	10.05	128.55

[1]O&M = operations and maintenance.

A dispatchable source of electricity is one whose power output can be adjusted to respond to electric grid demand. Unlike conventional electricity sources, wind and solar are non-dispatchable. Grid parity is achieved when the LCOE of an alternative energy source is less than or equal to the cost the consumer would pay a utility for electricity.

Different power plant types have different dispatch times, i.e., different intervals for ramp up or shut down in response to grid demand. Hydroelectric plants can be dispatched within about 20 s while natural gas plants can be dispatched within minutes. Coal, biomass, and nuclear plants can be dispatched within hours. However, for safety reasons, nuclear plants must be able to shut down within seconds. Dispatching is important because grid demand needs to be perfectly balanced with electricity supply or else blackouts occur.

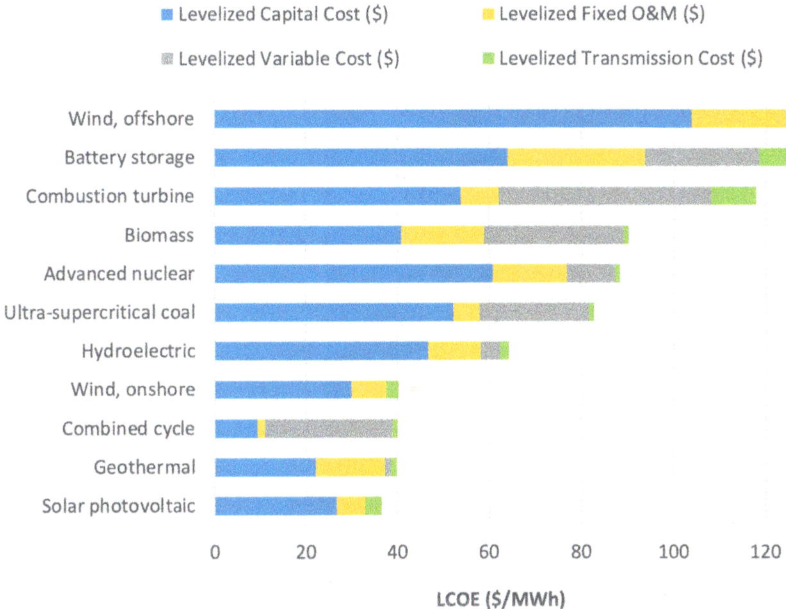

Figure 3.23: LCOE for different power plant types by cost category in 2021 dollars: capital, fixed O&M, variable O&M, and transmission cost. Source: U.S. Energy Information Administration, Annual Energy Outlook 2022.

The NGCC plant has very low capital cost and has among the lowest LCOE of most plant types. Offshore wind, on the other hand, is expensive to build and operate and therefore has a very high LCOE.

Grid demand must be perfectly balanced by supply of electricity into the grid. Any imbalance results in brownouts and blackouts. As a result, it is critical that electricity sources be dispatchable, i.e., able to ramp up and down in response to changing demand.

Electricity demand varies throughout the day, usually with a peak in the morning as people prepare to go to work, and another peak in the evening as people return home and use electricity for cooking, lighting, etc. A typical daily demand curve is shown in Figure 3.24. Demand can also vary by day of the week, as most people stay home on the weekends, and by time of year, e.g., with peaks in the summer due to increased use of air conditioning.

The duck curve is a plot of electricity production over the course of a day which shows the difference between electricity demand and solar power supply, as shown in Figure 3.25. The plot is shaped like a duck, hence its name. Utilities need to ramp up

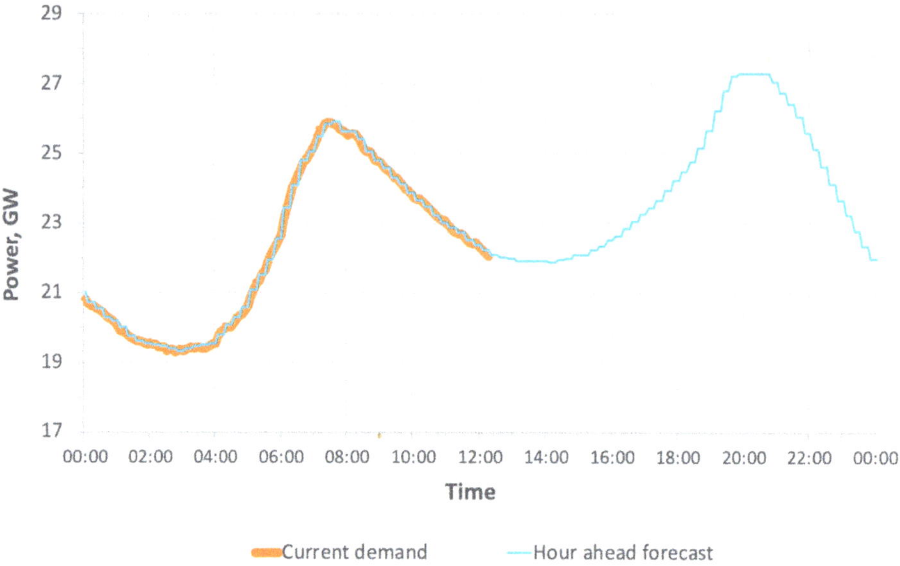

Figure 3.24: Demand curve and forecast for a typical day in March in California showing two demand peaks (captured on March 26, 2018) [44].

electricity production from other sources as the sun sets to compensate for solar power dropping. During the day, there can be excess production of solar power above electricity demand. This is called overgeneration and can lead to curtailment of solar power. Overgeneration can be mitigated by pairing solar power generation with storage. Another way to mitigate excess solar power production is through demand management. This includes switching to more efficient appliances and lighting systems to compensate for increased power demand at sunset, and implementing real-time pricing where consumers can pay lower prices during low demand periods to encourage less power use during peak demand.

Integration of high percentages of renewables like wind and solar into the grid is difficult due to their intermittent nature. A high share of renewables can lead to grid instability and unreliability. Integration of renewables can be facilitated by having a flexible electric system and implementing demand side management. Flexibility involves innovating coal and gas power plants to quickly ramp up to quickly respond to increased loads and adding storage to compensate for wind and solar variability [45].

3.4 Base Load Versus Peak Load

Base load is the minimum amount of electricity demand over a specified period, say 24 h. A power plant that supplies base load power needs to run continuously and can take days to come online or shut down. Base load plants are typically coal or nuclear.

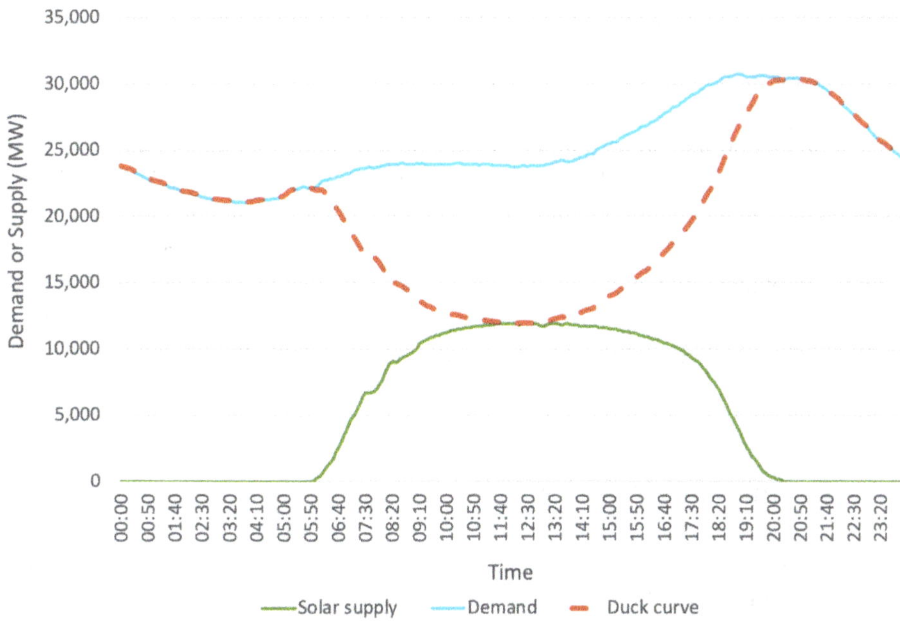

Figure 3.25: Demand, solar supply, and difference between the two (duck curve) for CAISO on June 17, 2020 [44].

Peak load refers to a spike in demand over a short period. Power plants that supply peak power are called peaker plants and they generally run only when there is a spike in demand. Peaker plants are typically natural gas. Diesel plants can be peaker plants as well though they are much more expensive to run than natural gas plants. In Figure 3.24, the base load is 19.5 GW while the two peak loads are 26 and 27.5 GW.

Chapter 4
CO_2 Capture and Storage

Development and industrialization have historically been driven by fossil fuels such as coal and oil, the use of which leads to emission of greenhouse gases (GHG) such as carbon dioxide (CO_2) and methane. In recent times, GHG have been tied to climate change and as such efforts are being made to incorporate more renewable energy such as solar and wind into the energy mix. This chapter explains the technical aspects of CO_2 capture and storage. It also compares CO_2 emissions by different energy sources for power generation and examines the cost added by CO_2 capture during power generation.

4.1 CO_2 Emissions in Africa

Africa accounted for only 4% of global CO_2 emissions in 2024 as shown in Figure 4.1 [46]. Thus, minimizing CO_2 emissions should not be a major focus of development work. It is more important to tackle the issue of insufficient electricity. Nonetheless, CO_2 emissions will be addressed here so that development workers may be armed with appropriate information if addressing CO_2 emissions becomes part of a project.

A good approach to addressing CO_2 emissions in development projects would be to study the potential impacts of climate change and include appropriate preparations in the project plans. For instance, in some cases, increased rainfall is expected as a result of climate change. In this case, it is important to examine the potential effects of increased rainfall on hydropower and the associated reservoirs and waterways. Preventing CO_2 emissions altogether is difficult since fuel combustion is necessary for a lot of electricity generation sources. Solar and wind power are the exceptions but their intermittent nature means that they cannot be the only electricity source on the grid especially in a development context.

4.2 CO_2 Emissions by Fuel Type

CO_2 is a gas that makes up 0.04% of air. Plants use CO_2 and sunlight to produce carbohydrates and oxygen. In electricity generation, CO_2 is produced when fuel is combusted. This is because fuels such as coal, biomass, and natural gas contain carbon, and the reaction of carbon with oxygen leads to production of heat and CO_2. Thus, it is unsurprising that CO_2 concentrations in the atmosphere have been on the increase since the start of the industrial revolution. It is thought that higher levels of CO_2 in the atmosphere will lead to higher temperatures and greater climate variability which

https://doi.org/10.1515/9783111643489-004

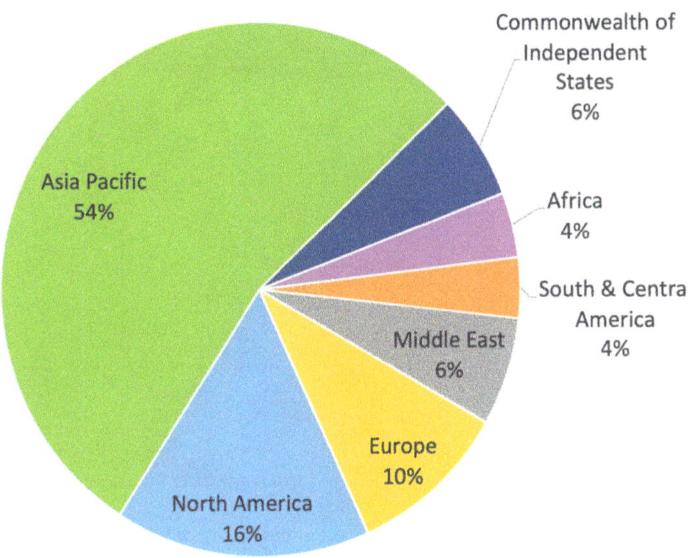

Figure 4.1: Percentage carbon emissions by region of the world in 2024 [46].

could have severe economic impacts [47]. As such, reducing CO_2 emissions is becoming a priority especially in developed countries.

Carbon capture and storage (CCS) is the process of collecting CO_2 to prevent its release into the atmosphere and storing it in geologic formations. Capturing CO_2 requires input of energy. This means capturing CO_2 adds to the cost of electricity generation. The process can be cost prohibitive. Converting CO_2 to other compounds is essentially reversing the generation of electricity since energy must be put back into CO_2.

To carry out a fair assessment of CO_2 emissions from various electricity fuel sources, a life cycle analysis must be conducted. This requires studying every step of the mining, harvesting and manufacture, transportation, and combustion of the fuels and other relevant materials and equipment. As expected, fossil fuels are the highest emitters with coal at 1,041 tons of CO_2 equivalent/GWh, followed by oil at 875 tons of CO_2 equivalent/GWh and natural gas at 622 tons of CO_2/GWh [48]. Wind and geothermal are on the low end with 15 and 14 tons of CO_2/GWh, respectively. CO_2 emissions for various fuels are shown in Figure 4.2 in tons of CO_2 equivalent per GW-hour.

4.3 How CO_2 Is Captured

The state-of-the art method of CO_2 capture is through the use of chemicals called amines. The process is illustrated in Figure 4.3. Amines react with CO_2 to form an amine-CO_2 complex which can be broken down back to CO_2 and amine simply by

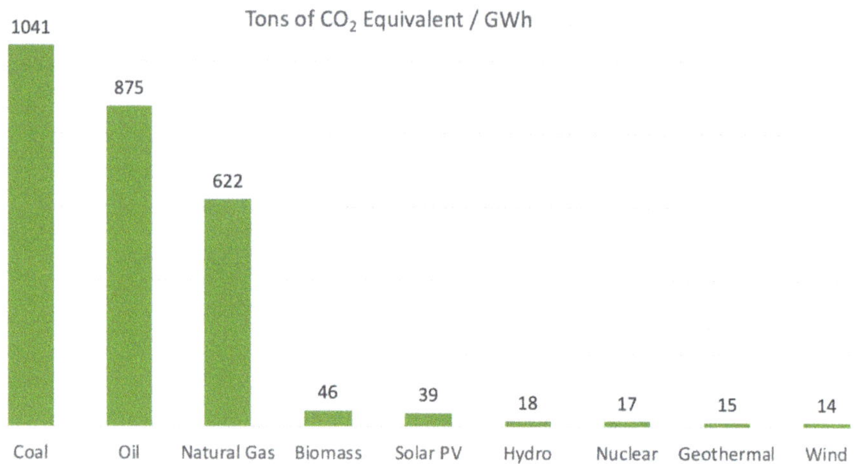

Figure 4.2: Tons of CO_2 equivalent emitted per GWh for life cycle of various electricity sources.

heating. In the absorber unit, CO_2 from flue gas combines with amine to form a complex. The complex is sent to a stripper unit which uses heat to strip CO_2 from the amine. The stripper operates at 120 °C and 200 kPa (29 psi). The CO_2 is compressed and sent to storage while the amine is recycled and sent back to the absorber to capture more CO_2.

CO_2 capture using amines has several disadvantages. The process requires a lot of energy to heat the amine-CO_2 complex and compress the captured CO_2. The energy requirements for the stripper are about 1.1 kWh/kg CO_2 [49]. Since the CO_2 in flue gas is usually present at low concentrations, around 5%, transfer into the amine is not very effective. This also results in the need for larger capture facilities. Amines also tend to degrade over time due to the presence of oxygen and other impurities, meaning make-up amine must be added despite the amine being recycled.

CO_2 can be captured before combustion or after combustion of fuels. For pre-combustion capture from coal, coal is converted to CO_2 and hydrogen through the gasification process. CO_2 is then captured and the hydrogen combusted to generate heat for conversion of water to steam. Oxycombustion is a pre-combustion capture method in which combustion is carried out in pure oxygen rather than air. The exhaust gas from combustion, i.e., flue gas, is mostly steam and CO_2. The mixture can be separated by cooling to remove water.

In post-combustion capture, coal is burned to generate electricity and CO_2 from the combustion process is then captured before the flue gas is vented. Flue gas from coal combustion contains CO_2, water vapor, SO_x, and NO_x. Power plants can be retrofitted to include post-combustion CO_2 capture more easily than pre-combustion. CO_2 capture can add 33–80% to the LCOE of coal plants, depending on whether oxycombustion, pre-combustion, or post-combustion is used, as shown in Figure 4.4 [50].

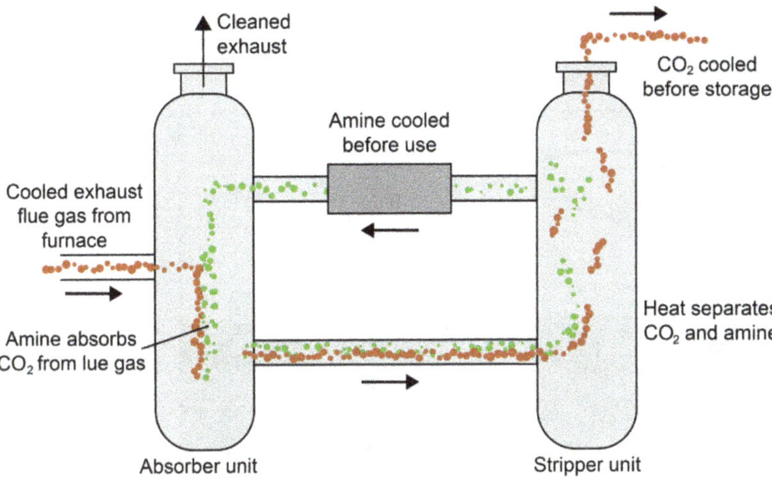

Figure 4.3: Carbon capture using amine solvents.

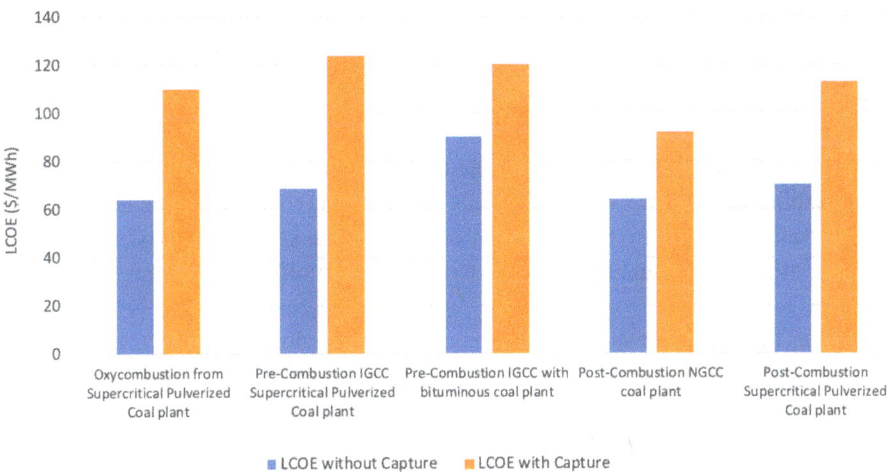

Figure 4.4: LCOE for coal plants with and without CO_2 capture [50].

4.4 Enhanced Oil Recovery

Captured CO_2 can be used for enhanced oil recovery (EOR) which generates revenue and stores CO_2 simultaneously. In this process, CO_2 is pumped into a depleted oil reservoir. The CO_2 pushes the oil through the reservoir to the producing well. It also mixes with the remaining oil, reducing its viscosity and making it easier to pump out. This increases the oil recovery from the reservoir by about 17%. Most of the CO_2 remains underground. Thus, EOR can be a profitable method of sequestering CO_2. The CO_2 EOR process is illustrated in Figure 4.5.

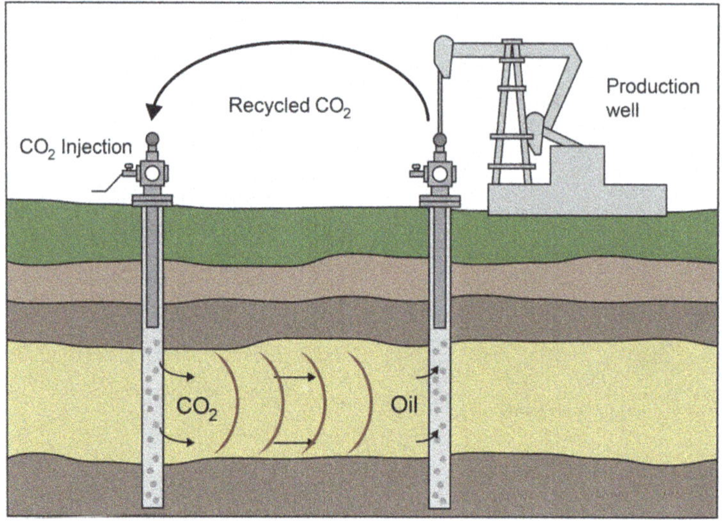

Figure 4.5: Enhanced oil recovery using CO_2 injection.

Aside from oil and gas reservoirs, storage of CO_2 can be done in geologic formations such as saline formations and coal seams. Injection in coal bed seams is called enhanced coal bed methane recovery. CO_2 adsorbs to the coal seam, driving coal bed methane off. The methane can be collected and sold to offset the cost of CO_2 sequestration.

In conclusion, CCS should not be a major concern in development work. It usually adds significantly to the cost of electricity generation, which would also increase the electricity tariffs paid by the end user. Instead, steps should be taken to mitigate any expected effects of climate change.

Chapter 5
The Energy–Water Nexus

This chapter examines the interrelationship between energy and water. Water is used in the production of electricity, so when choosing a power generation method, it is important to ensure that the local environment can supply enough water to support electricity production. This chapter will cover water needs for different electricity generation methods.

The energy–water nexus is the interconnection between water and the production of energy, whether that is electricity, or the extraction, processing, and transportation of energy. One example is the use of water for growing and processing of biomass. Energy is, likewise, used for extraction of water from water sources, water treatment, transportation, wastewater collection and treatment, and disposal of treated wastewater. The energy–water nexus is illustrated in Figure 5.1. For every energy source selected, the corresponding water use must be considered as well as the availability of necessary water amounts in that region.

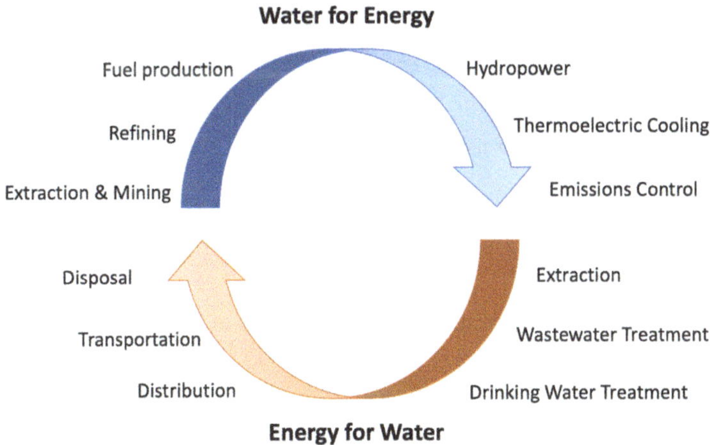

Figure 5.1: Energy–water nexus.

Across Africa in 2015, about 82% of fresh water is consumed by the agriculture sector, followed by 13% by municipalities and 5% by industry [51]. As electricity generation increases in Africa, so too will demand for fresh water for the energy sector. Fresh water use from aquifers, lakes, and other water bodies should be monitored to ensure that it is not higher than the replenishment rate from rainfall. Annual freshwater withdrawal in African countries is shown in Figure 5.2.

https://doi.org/10.1515/9783111643489-005

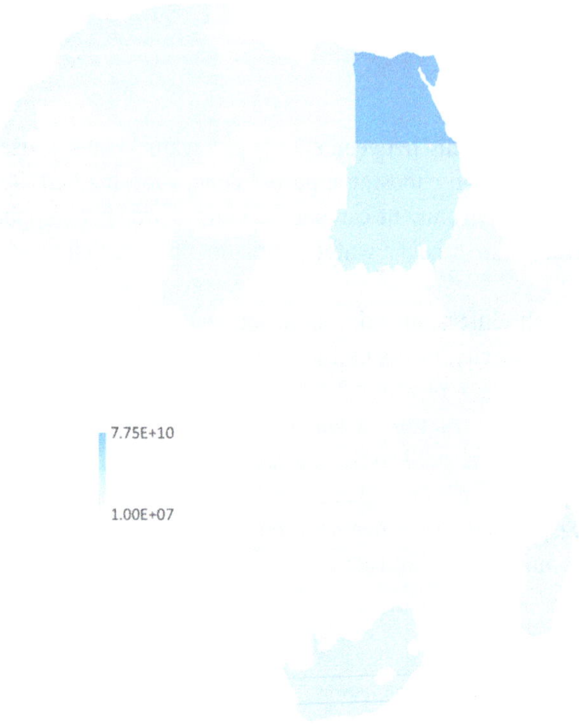

Figure 5.2: Annual freshwater withdrawals, total (billion cubic meters) in African countries in 2021 [52].

5.1 Energy Use for Water

Energy is used to extract water from water sources, treat water, distribute it to the consumer, collect the wastewater, treat the wastewater, and then either recycle the wastewater to the water treatment facility, or dispose of the treated wastewater in the water source. Energy uses for water are illustrated in Figure 5.3.

Energy consumption varies depending on the treatment method. Water treatment methods include desalination, chlorination, reverse osmosis, and membrane filtration. Figure 5.4 shows energy use by various forms of water treatment. Of course, treatment choices will first be determined by what needs to be removed from the water. For instance, to remove bacteria from water, ultraviolet light, sodium hypochlorite, reverse osmosis, and nanofiltration can be used. All else being equal, ultraviolet light uses the least amount of energy, 0.12 kWh/m^3 of treated water, and should be the top choice if energy use is the only consideration.

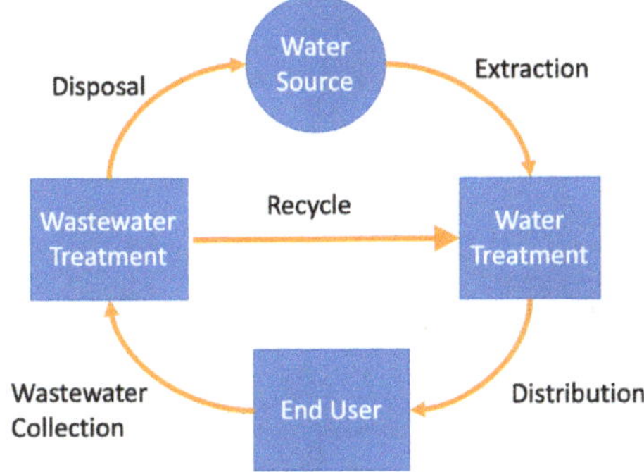

Figure 5.3: Stages of water system electricity needs.

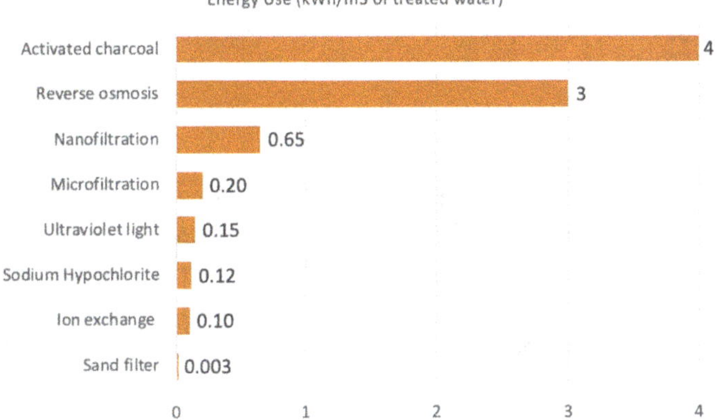

Figure 5.4: Energy use for various water treatment types [53–58].

5.2 Water Use in Power Generation

In a typical thermal power plant, water is used to generate steam which turns the blades of a turbine. Afterward, the steam condenses to hot water which is cooled by cold water. The cooled water is reused to generate steam and the cycle continues. In some cases, the cooling water is returned to the body of water it was taken from after it is sufficiently cooled itself. In other cases, the cooling water is cooled and reused. Water is also used to capture pollutants and dispose of waste such as ash in a coal plant. Uses for water in a thermal power plant are illustrated in Figure 5.5.

Water use can be measured in the form of water withdrawal or water consumption. Water withdrawal is the amount of water that a plant draws from a water source such as a lake, river, aquifer, ocean, or municipal system. Water that is consumed is no longer available and is not directly returned to a water source, i.e., water that is lost to evaporation.

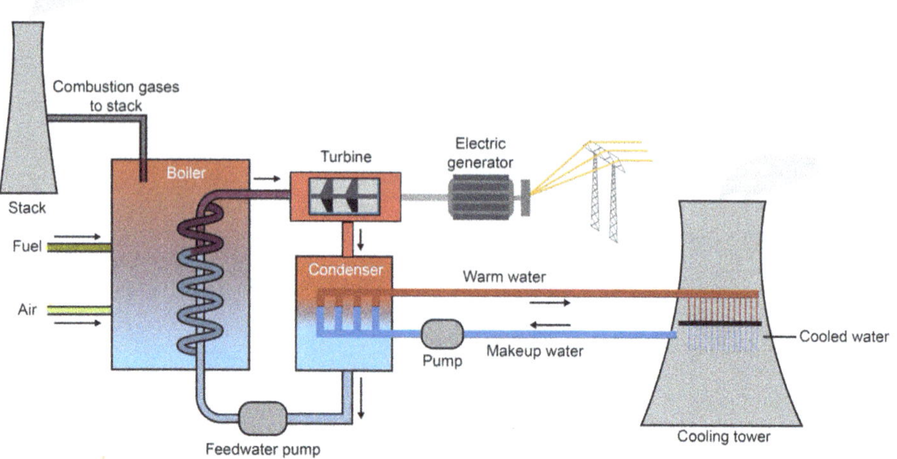

Figure 5.5: Water use at a typical thermal power plant.

Different power sources have different water usage needs. Figure 5.6 shows water consumption for electricity generation from different fuel sources. As expected, hydropower consumes the most water at about 17,000 L/MWh of electricity generated. Concentrated solar power, natural gas combined cycle, coal, nuclear, and biomass all use several thousand liters of water per MWh. Solar PV and wind use virtually no water.

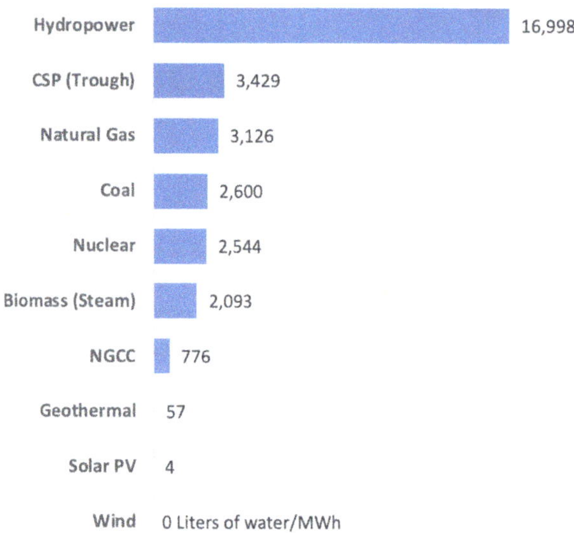

Figure 5.6: Water consumption for different electricity generation sources in liters of water per MWh [59].

Thermal power plants can have different water cooling configurations namely once-through, recirculating, and dry-cooled. Dry-cooled systems use air instead of water to cool steam coming from the turbine. While this cooling method uses no water, it does lead to lower efficiency since water is a more effective heat conductor than air.

In a closed-loop system, or recirculating system, hot water that has been used to condense steam is sent to a cooling tower which is an open-air tank. There, the water is exposed to air and cools down. The water is then sent back to the condenser to cool down steam from the turbine. In the cooling tower, some water is lost to evaporation and must be replaced by pulling more water from a water source.

In a once-through system, water is extracted from a source such as a lake or river, used to condense steam from the turbine, then sent back to the water source. Since no cooling tower is used in this system, the water returned to the source can be at temperatures about 37 °C higher that the water in the source. This can harm fish and other organisms in the water and can lead to algal blooms.

Both once-through and recirculating systems can circulate cooling water through onsite reservoirs called cooling ponds. Like cooling towers, cooling ponds lose water to evaporation.

Figure 5.7 shows water withdrawal and consumption for different cooling configurations for a coal power plant. Recirculation using a cooling tower withdraws the least amount of water at 2,298 L/MWh but has the highest water consumption at 1,529 L/MWh. Once-through with a cooling pond withdraws 115,338 L/MWh and consumes 1,287 L/MWh. A recirculating system with a cooling pond withdraws 133,769 L/MWh and consumes 1,393 L/MWh. In choosing the cooling configuration for a thermal power plant, water availability in the region should be taken into consideration.

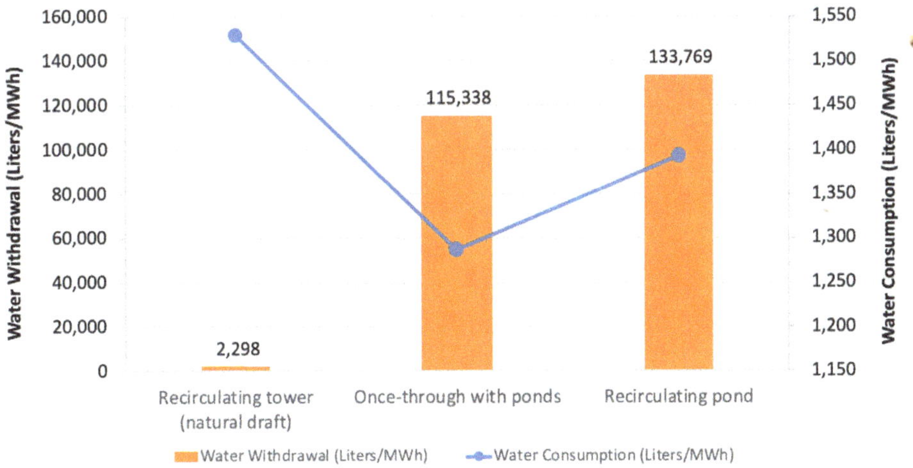

Figure 5.7: Water withdrawal and consumption for coal power plants with different cooling configurations [60].

Droughts can lead to power plants reducing their electricity output [61]. Heat waves can also cause power plants to reduce output when their electricity is most needed. This is due to temperature limitations on the water being returned to the water source in once-through systems to prevent harm to aquatic life. Even dry-cooled plants are affected by heat waves since hot air is less efficient at cooling, leading to a drop in plant efficiency.

Since energy and water are interconnected, energy efficiency measures also save on water usage. Efficiency measures can be taken to reduce the amount of water used for electricity generation. For instance, for coal power generation, using more efficient technologies like supercritical power generation reduces the amount of water used. Implementing monitoring of water use through the use of sensors can help to identify leaks. Using closed-loop systems in power plants where water is recycled results in less withdrawal of water from water bodies than open-loop systems. In choosing the cooling configuration for a thermal power plant, water availability in the region should be taken into consideration.

Chapter 6
Energy Policy

This chapter delves into energy policy because while energy technology is important, infrastructure investments are more likely to be sustainable if the appropriate policies are in place to encourage and incentivize sustainable investments in energy infrastructure.

Due to the cross-cutting nature of energy and its impact on virtually every other sector, it is important for every country to have a strong energy policy to ensure energy security. Energy policy governs the production, distribution, and consumption of energy to ensure sufficient energy supply to drive the economy of a country. Almost every country has an energy roadmap which details its strategy for achieving its energy goals. The roadmap may be used to prioritize some energy projects over others.

Policy should be set, monitored, and altered as needed. Policy should be enforced as should penalties for failing to follow set policy, or else the policy is useless. Since policy always has benefits and drawbacks, energy policy should strive to strike a balance among all three pillars of sustainability. Energy policies should secure energy supply while ensuring fair pricing and minimizing damage to the environment. Since information is essential to decision-making, policies should also encourage collection and analysis of energy data.

Energy policy encompasses policies, laws, and regulations. A policy outlines the goals of the government and how they will be achieved. While a policy is not a law, it identifies new laws that need to be passed for the goals to be achieved. Laws are passed by the legislative body of the country, and they remain in place even when a new government takes power. Regulations are rules that provide detailed direction to an implementing agency or other actor on how to implement laws.

The typical energy sector players include the ministry of energy, the rural electrification government agency, the electric utility, a regulatory agency, and private sector players such as oil and gas companies. The regulatory agency is usually an independent body with the authority to regulate energy sector activities including electricity transmission, sales of electricity, and oil and gas. The electric utility may comprise generation, transmission, and distribution. In some cases, there may be two or more utilities focused on generation, transmission, or distribution, or two out of the three. This is referred to as vertical unbundling.

An important part of energy policy is ensuring that the private sector is encouraged to participate in the energy sector. This may include opening up the sector to independent power producers (IPPs) and introducing policies to ensure that electricity produced by the IPPs will have a guaranteed off-taker, i.e., a guaranteed buyer for the electricity such as the utility.

https://doi.org/10.1515/9783111643489-006

6.1 Energy Policy Reform

Energy policy reform is viewed as more and more important in development work in ensuring the sustainability of energy infrastructure projects. While policy reform may seem like a good idea, it can cause more damage if it is done in too short a timeframe, without follow up, and without taking cultural and local context into consideration. Strong policies can only be upheld by strong institutions, and it takes years to strengthen institutions. Policy reform helps to improve sector governance, strengthen the regulatory role, and improve performance of the utility.

Energy policy reform can involve making changes to policies, laws, and regulations or creating new ones. This sort of reform is typically done with the ministry of energy or the regulatory body. Policy reform may also involve creating new institutions or changing existing ones to perform new roles. This may require provision of capacity development to ensure that the new role is understood and performed correctly.

6.2 Renewable Energy

Solar and wind energy policies in emerging economies are generally weak or nonexistent due to the fact that their proliferation is recent. New policies should encourage adoption and deployment of these two technologies, particularly solar energy as it is more affordable for distributed energy use. However, solar panels are not the answer to every problem.

Policies that encourage adoption of renewable energy include allowing individuals and microgrids to sell excess power back into the grid. This is done by payment of a feed-in tariff (FiT) which is a fixed price paid to renewable energy producers for a set period of time for injecting their electricity into the grid. For countries with an electricity deficit, this could go a long way toward reducing that deficit. On the other hand, there is no guarantee that electricity will be available from the grid when the solar producer needs to buy electricity at night. Thus, this would work well if the solar producer has battery storage as well. With FiT, two electric meters are needed as the utility pays one price for generation, and another, usually lower price, for consumption. Net metering is another way in which the utility can compensate renewable energy producers for excess electricity. Only one meter is involved and the same price is used for generation and consumption. The utility provides a credit for the amount of electricity sent back into the grid. The utility may cap the amount they will credit. Above the cap, no further compensation will be provided.

Another way to encourage renewable energy adoption is to enact policies that guarantee renewable energy producers a market for their electricity. Similar policies have been set in some US states and they have contributed to increasing adoption of wind and solar power.

Renewable energy adoption can be encouraged by providing rebates for purchasing renewable energy equipment. This reduces the cost of the system and makes the equipment more affordable. Similarly, tax incentives can be offered for buying and installing renewable energy systems. These tax incentives can take different forms. They can offer cash or tax credits to deduct a portion of the cost of the system.

Subsidized loans can be offered for renewable energy systems, preferably with reduced interest rates. Accelerated depreciation can also be offered to reduce the tax burden for the renewable energy provider.

Disposal of solar panels, batteries, and wind turbines is a relatively recent problem to deal with. Policies need to be created to mandate proper disposal of this equipment, as well as to provide guidance on how the disposal should be done.

6.3 Carbon Trading

Carbon markets are tools to mitigate greenhouse gas (GHG) emissions by allowing trading of carbon offsets. A carbon offset, or carbon credit, represents the reduction, avoidance, or sequestration of 1 metric ton of CO_2 or the GHG equivalent. Businesses with climate change mitigating projects such as solar power projects can sell emissions reductions to other businesses, particularly in industrialized countries. The Clean Development Mechanism (CDM) through the Kyoto Protocol and the Sustainable Development Mechanism (SDM) through the Paris Climate Agreement allow industrialized countries with GHG reduction requirements to invest in GHG-reducing projects in developing countries. Voluntary carbon markets include any carbon offset trading that is not part of a compliance marketplace but carried out on a voluntary basis.

Carbon offsets can take the form of voluntary emission reductions (VERs) on the voluntary markets or certified emission reductions (CERs) on the compliance markets. Since Africa emits only 4% of the world's carbon dioxide, countries in Africa can take advantage of this and sell their carbon credits to customers looking to reduce their carbon footprint. However, policies need to be put in place to encourage carbon trading.

Only 261 out of 8,814 projects, or 3%, of carbon trading through the Clean Development Mechanism (CDM) came from African countries in 2016 [62]. Africa's share of the voluntary carbon market was even smaller at 1%.

Policies that could increase carbon trading in Africa include putting a price on carbon, which most African countries have not done yet, establishing carbon markets in Africa, and increasing awareness of and capacity for the legal and regulatory aspects of carbon trading.

6.4 Energy Efficiency

Energy efficiency is the avoidance or reduction of energy waste. Energy efficiency leads to fewer resources such as fuel and water being used in the generation of energy, as well as fewer emissions per unit of electricity generated. There should be policy alignment between energy and water flows since increased electricity generated leads to increased water use. As explained in Chapter 3, electricity supply and demand must be perfectly balanced on the grid or else blackouts occur. Implementing energy efficiency measures is critical in situations where electricity demand far outweighs supply.

Energy efficiency policies can address either the production or consumption of electricity. The range of energy efficiency measures from system to individual is shown in Figure 6.1. On the production side, the goal is to produce more energy using the same amount of resources. On the consumer side, the goal is to reduce peak demand or total energy demand. The consumer side is called demand side management, and it is addressed in the paragraphs that follow.

System Improvements ⟷ Behavioral Improvements

Policy	Infrastructure	Appliances	Demand side Management	Education
Energy efficiency standards	Optimization	Energy audits	Lower prices for off-peak energy use	Customer awareness campaigns
Building energy codes	System upgrades	Energy efficient devices	Pricing tier	Employee training
	Smart meters			

Figure 6.1: Energy efficiency measures.

Electricity is most expensive during peak demand periods. However, the cost of electricity is generally constant for the end user. Tariffs can be set lower for off-peak electricity use than during peak demand to reduce electricity use during peak demand periods. Smart meters are necessary to track usage during peak- and off-peak periods.

Reducing total electricity use by the consumer can be done by creating tiers for total electricity use and charging more for tiers that use more electricity. The consumer pays one price for the first tier of electricity, say up to 300 kWh, then pays a higher price for the next tier of electricity use, say from 301 to 500 kWh, and so on.

Another way to encourage energy efficiency is to provide incentives such as rebates for the consumer to purchase energy audits and energy efficient devices such as LED lightbulbs. Energy efficiency standards should be created for appliances and equipment. Similarly, building energy codes can be put in place to make their energy use more efficient. An energy audit of a home or commercial building shows how much energy is used in the building, what it is used for, where energy is being lost, and where efficiencies can be gained.

6.5 Cybersecurity

Grid cybersecurity is the art of protecting the grid from unauthorized access. Cyberattacks can cripple an electric grid causing prolonged power outages, so protecting the grid is an energy security issue. Attacks on the grid also affect other sectors such as communications, health, and water. Cyberattacks on the grid are possible due to internet-enabled devices and smart meters being incorporated into the grid to create a smart grid, i.e., a grid that allows the flow of operational information.

Cybersecurity policy should address both deliberate attacks and vulnerabilities caused by natural disasters, user error, and equipment failure [63].

Cybersecurity standards should be created, although they are not sufficient by themselves for tackling cybersecurity risks. The standards could address [64]
- Critical cyber asset identification;
- Security management controls;
- Personnel and training;
- Electronic security;
- Physical security of critical cyber assets;
- Systems security;
- Incident reporting and response planning; and
- Recovery plans for critical cyber assets.

Protocols must be put in place for response to and recovery from successful attacks. Additionally, incentives could be created to encourage continuous adaptation to rapidly evolving cyber threats. An industry-led institute could be created to provide training and cybersecurity performance criteria and advance risk management practices throughout the energy sector. This new institute could also assess energy facilities and provide ratings on the readiness of the facilities to prevent and respond to cyberattacks. The institute and government can provide certifications on readiness levels of grid facilities and equipment. Finally, the institute could be a central point for information sharing on cyber incidents to make the industry aware of types and frequency of cyber incidents.

Government could create a cybersecurity task force with members from the entire energy sector. Additionally, government could create reporting standards and requirements for cyber events, including setting actions to be taken by government players in preparation for and response to cyber events.

6.6 Regional Energy Policy

A power pool is a mechanism for electric utilities to trade electricity. A power pool creates a common framework and rules for pooling energy resources, encouraging power exchanges, and reducing power supply costs. The advantage of a power pool is that it creates a common market for electricity companies in the member countries. It provides enhanced energy security by ensuring electricity supply for the members, as well as a platform for coordinated expansion of power generation and transmission. In most cases, there will be a few members that sell electricity into the pool while the other members buy from the pool.

Africa has five different regional power pools namely the West Africa Power Pool (WAPP), the Southern African Power Pool (SAPP), the North African Power Pool, also called Comité Maghrébin de l'Electricité (COMELEC), the Central African Power Pool, also called Pool Energetique De L'Afrique Centrale (PEAC), and the Eastern Africa Power Pool (EAPP).

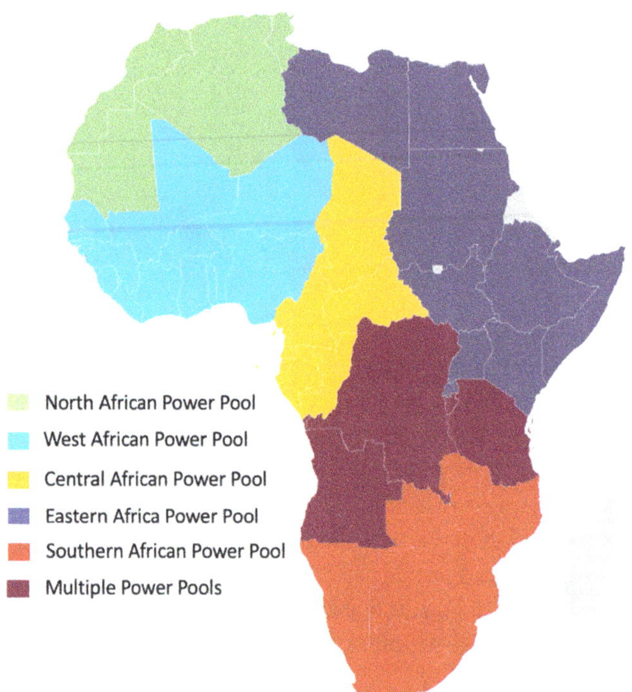

Figure 6.2: The five regional power pools in Africa. Note: The Democratic Republic of Congo belongs to the Southern African Power Pool, the Eastern Africa Power Pool, and the Central African Power Pool; Tanzania belongs to the Eastern Africa Power Pool and the Southern African Power Pool; Angola belongs to the Southern African Power Pool and the Central African Power Pool; and Burundi belongs to the Central African Power Pool and the Eastern African Power Pool.

WAPP has 14 members in West Africa and they are Nigeria, Mali, Cote d'Ivoire, Guinea, Guinea Bissau, The Gambia, Senegal, Liberia, Benin, Togo, Ghana, Sierra Leone, Niger, and Burkina Faso [65]. SAPP has 12 member countries namely Zambia, Zimbabwe, South Africa, Lesotho, eSwatini (formerly Swaziland), Mozambique, Malawi, Namibia, Angola, Botswana, Tanzania, and the Democratic Republic of Congo [66]. COMELEC has five members countries namely Algeria, Morocco, Libya, Tunisia, and Mauritania. PEAC covers 10 members countries. They are Angola, Burundi, Cameroon, Congo, Democratic Republic of Congo, Central African Republic, Gabon, Equatorial Guinea, Sao Tomé and Principe, and Chad [67]. EAPP has 13 members. They are Kenya, Tanzania, Uganda, Djibouti, Egypt, Burundi, Rwanda, Somalia, Sudan, South Sudan, Ethiopia, Libya, and the Democratic Republic of Congo [68]. The distribution of the power pools in Africa is shown in Figure 6.2. Note that the Democratic Republic of Congo belongs to the Southern African Power Pool, the Eastern Africa Power Pool, and the Central African Power Pool; Tanzania belongs to the Eastern Africa Power Pool and the Southern African Power Pool; Angola belongs to the Southern African Power Pool and the Central African Power Pool; and Burundi belongs to the Central African Power Pool and the Eastern African Power Pool.

Electric utilities, not governments, are the members of the power pools. This explains why some countries are members of more than one power pool. The different utilities in these countries belong to different power pools, sometimes for geographic convenience. Over 62% of electricity generated in SAPP in 2016 came from coal, followed by hydropower at 21%, as shown in Figure 6.3. This is because most electricity is supplied by South Africa which gets most of its electricity from coal.

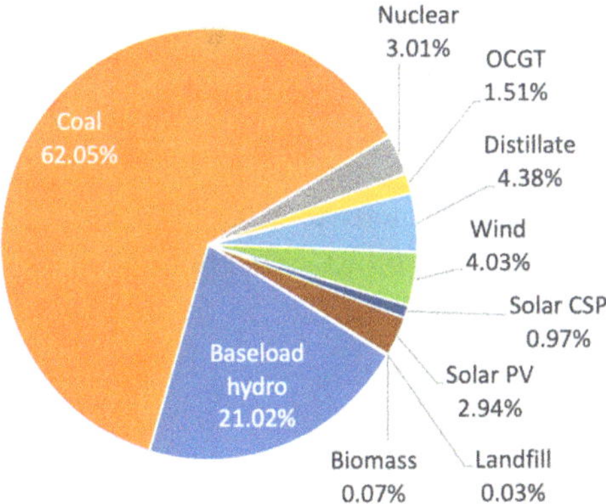

Figure 6.3: SAPP power generation sources in 2016. OCGT is open cycle gas turbine, CSP is concentrated solar power, and PV is photovoltaic [69].

Figure 6.4 shows electricity imports and exports within WAPP in 2020. Nigeria, Ghana, and Cote d'Ivoire are the electricity exporters with Nigeria exporting the most elec-

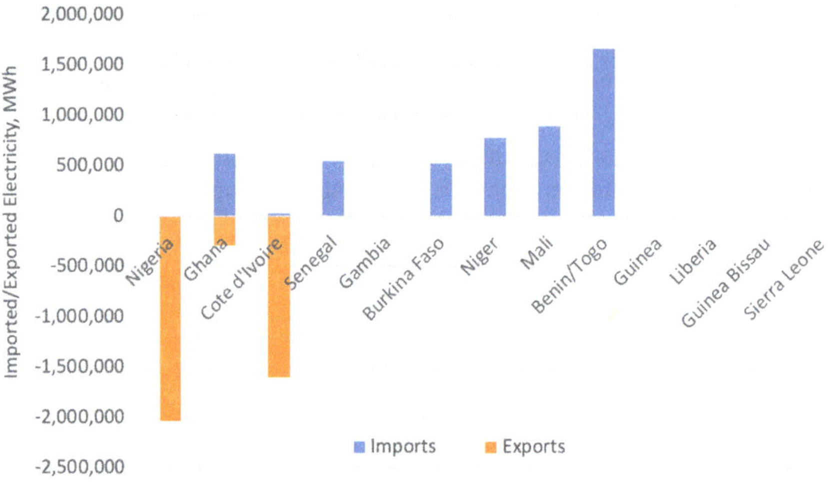

Figure 6.4: WAPP electricity imports and exports by country in 2020 [70].

tricity at 2 GWh. The countries that import electricity are Ghana, Cote d'Ivoire, Senegal, Burkina Faso, Niger, Mali, Benin, and Togo. The remaining countries consume all the electricity produced within their borders.

Power pools work best if grid interconnections are well-developed, and there is a robust legal framework for cross-border import and export of electricity. An appropriate dispute resolution mechanism is also necessary for the power pool to work.

A key challenge for WAPP is its inability to finance big generation and transmission and distribution projects in more remote regions. EAPP is looking to expand and improve existing transmission and distribution infrastructure, as well as develop transmission interconnectors between member countries. The key focus for SAPP is increasing generation capacity and improving its interconnection network. CAPP is the least developed power pool in Africa with 75% of Central Africa lacking access to electricity. CAPP's challenges include a poor regional framework for electricity trading, weak regional regulations for dispute management, difficulty in attracting investments, and low interconnection between member countries [71].

6.7 Electric Utility

An electric utility is a commercial entity that is involved in generation, transmission, and/or distribution of electricity to a consumer. The electric utility is one of the key players in the energy sector. Many utilities in Africa are part of the government. Regulations should make it easy for the private sector to participate so that there is enough electricity supply. The utility may be vertically bundled, i.e., having generation, transmission, and distribution, or it could be unbundled with a separate utility for each function. Generation utilities own power plants while transmission and distribution utilities own the power lines, poles, and other equipment used to deliver electricity to the consumer. The relationship between generation, transmission, and distribution utilities is shown in Figure 6.5.

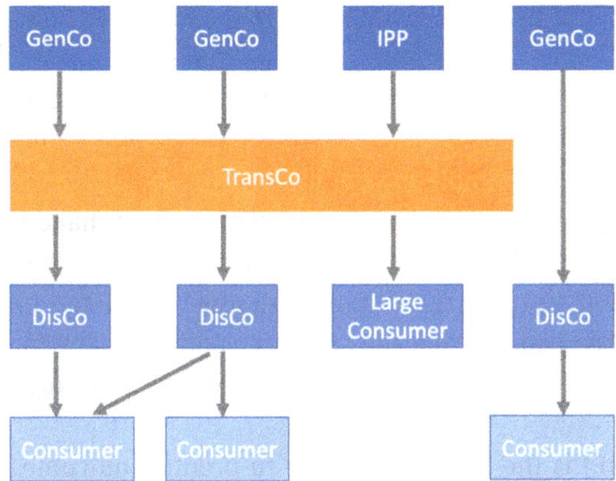

Figure 6.5: Relationship between power sector utilities. GenCo refers to power generation company; IPP refers to independent power producer; TransCo refers to transmission company; and DisCo refers to distribution company.

For vertically bundled utilities, it can be difficult to pinpoint what part of the utility the income and profits or losses stem from. Energy policy reforms typically strive to ensure the financial viability of the utility and separate the cash flows for different parts of the utility. The utility being profitable means that it can invest in maintenance and expansion of the grid and power plants. This leads in turn to the consumers being provided with good quality and reliable electricity service. Unbundling is not appropriate for every situation. It can cause more issues if it results in two or three new entities that are not financially viable.

The first step involves ensuring that the utility is charging a fair price for electricity, i.e., cost-reflective tariffs. This requires understanding the true cost of electricity generation, transmission, and distribution. Any subsidies that are available to consumers need to be considered as well. Subsidies are used to increase access to electricity. The government pays for some or all of the electricity consumed by rural or poor consumers. Alternatively, the subsidies may pay for connection costs rather than electricity consumption.

Electricity losses may be technical or nontechnical. Technical losses are encountered during voltage step-up and step-down, transmission, and distribution of electricity. Technical losses may be reduced by upgrading grid lines and equipment or increasing transmission voltage. Nontechnical losses are essentially electricity that is consumed but not billed for as the consumer is unknown and the amount of electricity consumed is uncertain. These losses may come from theft of electricity or misallocated electricity flows. Nontechnical losses may be mitigated by having a complete inventory of the utility's assets and monitoring their operations using a SCADA system. SCADA stands for Supervisory Control and Data Acquisition. It is an automation control system with a centralized control system used in industrial complexes. Operators can use it to gather data in real-time and make adjustments to grid and power plant operations.

Rural electrification is difficult to accomplish because rural areas are sparsely populated and typically have a lower average income than urban areas. Extending grid lines to rural areas is unlikely to be profitable for the utility. Anchor customers in rural areas are customers that use and can pay for enough electricity to make it worthwhile for the grid to be extended. These could be as big as factories or as small as wealthy individuals. The connection costs for the first lines to a new area can be expensive. Other consumers in the area benefit from the anchor customer paying for the first line as subsequent distribution lines and transformers are typically much cheaper.

Productive use of electricity (PUE) is the use of electricity to produce outcomes that can be measured in monetary terms [72]. PUE is often considered mandatory for sustainable rural electrification in development work. In the past, rural electrification efforts have focused on household lighting which has resulted in lower returns than expected. It has proven difficult to connect access to electricity to the alleviation of poverty [73]. Perhaps access to electricity should be its own goal and not tied to PUE to be considered necessary or a success. However, this needs to be balanced with the utility's need to be a commercially viable entity.

One way to improve the financial viability of the utility is to increase the customer base by increasing connection rates. Experience has shown that it is not enough to extend the grid and expect people to connect. Studies have to be carried out to understand the factors affecting connection rates, and these vary from one region to another due to different electrification rates, geographies, etc. In some cases, the process of requesting a new connection is complicated and can take a long time. This can dissuade potential customers from starting the process. High connection costs can be a deterrent but they can be overcome by the utility spreading the cost of new connections over the entire customer base or providing payment plans for new connections.

6.8 Privatization

Privatization is the transfer of public institutions to private control and/or ownership. Privatization has been touted as a way to improve the financial viability of the utility and improving the efficiency of its operations. The experience in Australia has been that the privatized utilities' operations have become less efficient as they have a monopoly due to the high cost of entry into the sector [74]. Privatization means the government has less insight into and less oversight of the utility's operations. Opponents of privatization argue that necessities like electricity should be kept as a public service rather than for profit. Only about 14% of utilities in Sub-Saharan Africa are privatized [75].

Privatization or Public–Private Partnership (PPP) is a partnership between government and the private sector to deliver services or projects traditionally delivered by government. It can help keep public debt under control by transferring debt to the private sector. PPPs range from service contracts to full on divestiture. The main types of privatization or private sector participation are service contracts; management contracts; lease contracts; concessions; Build–Operate–Transfer (BOT), Build–Own–Operate–Transfer (BOOT), or Build–Own–Operate (BOO) concessions; and divestiture. The defining features of each are summarized in Table 6.1.

In a service contract, the utility sources goods or services from the private sector such as meter readings. Ownership of and responsibility for assets remain with the government. Under management contract, a private contractor collects tariffs for the government and is in turn paid an agreed upon fee [76].

Under a lease contract, the private sector operates and maintains the asset but is not responsible for financing. The contractor has greater commercial risk than with a management contract. The contractor charges a fee to the consumer and gives a portion to the government as a lease fee. The lease fee is fixed and the operator takes on the risk of nonpayment from consumers. Under affermage, or lease-and-operate, the private contractor pays a lease fee and is responsible for maintaining and operating the infrastructure while the government is responsible for financing new infrastructure [77]. In an affermage, the operator fee is guaranteed and the government takes on the risk of nonpayment from consumers.

In a concession agreement, the concessionaire is responsible for all operation, maintenance, and upgrading of infrastructure for a set period, usually 25–30 years. The government retains ownership of the asset. The concessionaire obtains its revenue directly from the consumer and pays a lease fee to the government. The lease fee is usually put toward upgrading and expansion of the asset [78].

BOT, BOOT, and BOO concessions typically last from 25 to 30 years. Under BOO, ownership remains permanently with the private sector. The private sector operates and maintains the asset. In a BOOT project, the contractor finances, builds, and operates a discrete asset for a set period of time. Then ownership is transferred to the government. The contractor is paid by the government rather than from consumer tariffs usually through a Power Purchase Agreement. BOT projects are financed by the government but built and operated by the private sector. Ownership remains with the private sector until the concession period ends, and then asset ownership is transferred to the government.

In a divestiture or privatization, there is a permanent change of ownership to the private sector. The government retains some form of control over the privatized utility in the form of license granted to deliver a service to the public.

Table 6.1: Main types of private sector participation and their defining features [79].

	Service contracts	**Management contracts**	**Lease contracts**	**Concessions**	**BOT, BOOT, and BOO concessions**	**Divestiture/ privatization**
Scope	Different contracts for support services (e.g., meter reading and billing)	Management of entire operations or major components	Responsibility for management, operations, and specific renewals	Responsibility for all operations, financing, and execution of specific investments	Investment in and operation of specific major components (e.g., a transmission line)	Responsibility for all operations, financing, and execution of investments
Asset ownership	Public	Public	Public	Public/private	Public/ private	Private
Contract tenure	1–3 years	2–5 years	10–15 years	25–30 years	Varies	License for 25–30 years
O&M responsibility	Public	Private	Private	Private	Private	Private
Capital investment	Public	Public	Public	Private	Private	Private
Commercial risk	Public	Public	Shared	Private	Private	Private
Risk assumed by private sector	Minimal	Minimal/ moderate	Moderate	High	High	High

O&M, operation and maintenance.

For a PPP to be successful, there must be a clear framework defining how investors will be paid, how tariffs will be set, and how commercial losses will be mitigated. If a PPP is executed well, risks will be borne by the sectors that can best manage them. Some of these risks are cost-overruns during construction, unexpectedly high operating costs, and unexpectedly low income.

If privatization is chosen as a solution, the energy roadmap should be used as a guide and to create a definition of success. For concessionaires, there should be close cooperation with the local utilities [80].

Privatization should not be assumed as the correct solution in every case. It should be done if the utility has a large customer base that would be attractive to private sector players and if the necessary management capacity cannot be developed within the country. The historical and political context matters as well. Some countries have tried privatization with poor results and are therefore unwilling to try again. Privatization should not be chosen as a path forward if there is no government commitment. Incentives should be very closely tied to the outcomes that are a priority for the government such as increased connections. If increasing demand is enough for the concessionaire to make a profit, the concessionaire is unlikely to invest in upgrading infrastructure. It is also important to understand that there is no one right approach to privatization. Privatization is done differently in different parts of the world. For instance, in the United States, some airports are state-owned while others are more privatized. Context matters and should be taken into serious consideration.

Chapter 7
Framework for Sustainable Energy Development

This chapter introduces a framework for addressing environmental, social, and economic development issues in a sustainable way. In a typical development scenario, locals are not consulted on what their issues are. Instead, outsiders develop a solution to a problem that the local community has not necessarily complained about. Even when the local community has identified a problem, the solution is often developed without their input. Excluding the local community from the planning process often leads to unsustainable solutions which are abandoned soon after the end of the project as they do not fit the needs and context of the local community. A one-size-fits-all approach should be avoided and solutions should be adapted for local culture.

Development organizations are often reluctant to significantly involve locals in development projects. There is an assumption on the part of development workers that they are more informed and more experienced than the beneficiaries. These misguided attitudes are likely rooted in racial bias and colonial attitudes and must be eradicated [81–83]. This can be done by involving more people from developing nations in development institutions in more than just token numbers and restructuring the institutions to help them succeed in their roles. It would also be helpful to provide training to help development workers recognize their unconscious biases and provide incentives to reward workers who involve beneficiaries in planning and implementation. Africans in the diaspora are the ideal candidates for roles in development organizations since they understand both local context and western context which includes how the development organizations work. They can act as a bridge to local communities, helping to translate culture and context for their western counterparts.

The framework presented in this chapter is useful for providing appropriate solutions rather than solutions that we think are needed. Key items of note are the importance of cultural context and stakeholder input in the process. These help with identifying the appropriate solutions and with ensuring that the solution is sustainable. If the stakeholders are not involved in the process, they are unlikely to adopt the solution permanently or adopt it at all. It is also a sign of respect to involve the stakeholders since it is their community which will be impacted by the project. Imposition of a solution is detrimental to their adoption of it.

The framework for planning development projects is summarized in Figure 7.1. Each step is further described below. It should be noted that stakeholders should be engaged at every single step. They should be involved in identifying the problem at the beginning of the process, identifying and selecting solutions, implementing the selected solution, and reporting on the success of the implemented solution.

Monitoring and evaluation (M&E) is another key component that should be applied from the start and throughout the framework. M&E allows for setting up targets that tie to the overall goal in a specific and measurable way and identifying metrics to

https://doi.org/10.1515/9783111643489-007

be tracked so that the impact of the project can be measured. M&E helps to determine if a project is successful in achieving its goal. The identified metrics must be measured at the beginning of a project to establish a baseline, throughout the project to measure progress, and for several years after the project ends to measure sustainability and success.

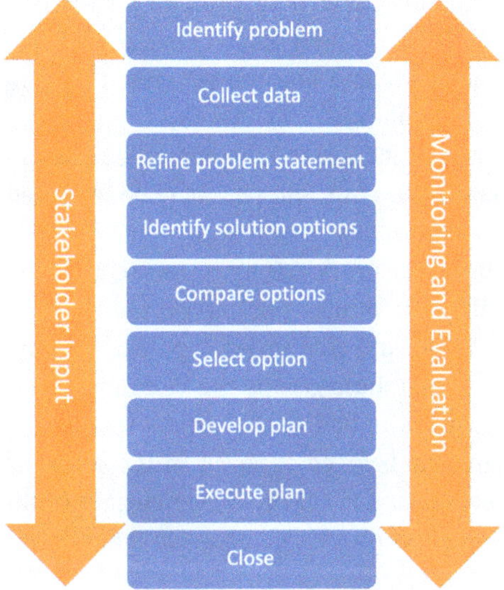

Figure 7.1: Framework for planning development projects.

7.1 Identify Problem

It is important to clearly identify and articulate a problem in order to identify solutions that will actually solve the problem. For instance, building a new power plant will not solve issues related to reliability of electricity as they are associated with the grid. Resources are wasted if the wrong problem is addressed. Even worse, addressing the wrong problem could end up being detrimental to the community by exacerbating the real problems. The local community know better than anyone what their issues are, so they should be consulted to help with identifying the problem. A decision needs to be made early in this step about how much weight the local stakeholders get in the decision-making versus the development organization. For instance, the local opinion could carry 60% of the weight and the development organization 40%. Either way, the various weights in decision-making need to be communicated early on to the beneficiaries.

This step could be called a root cause analysis. It leads to identifying the underlying issue rather than merely the symptoms. One way to get to the root cause is to keep asking "why" until the root cause is established. For instance, if someone is malnourished, it may seem like giving them money to buy more food is the answer. However, giving them money could end with them remaining malnourished but eating more expensive food. The problem could be that they do not understand nutrition, or it could be that nutritional food is not available in their area. You need to keep asking "why" until you get to the actual root problem. It is also important to consult the right local people during this step as having relevant, informed local people is invaluable.

Let us go back to the example of unreliable electricity. In this example, power interruptions are common and unannounced. This could stem from an insufficient electricity supply, in which case new power plants or power imports could solve the problem. However, it could also be due to an aging and poorly maintained grid, in which case upgrading the grid would be the right solution.

In this step, beneficiaries and all other stakeholders should be identified and engaged. This includes anyone who is impacted by the issue and who may be impacted by the solution.

The stakeholders can vary widely for an energy project. They could include electricity consumers in an area, the electric utility, the ministry of energy, the ministry of finance, nongovernmental organizations (NGOs) and civil society, and energy players in the private sector. The relationships between these groups can be complex and difficult to map. Having someone on the ground who has worked in the sector long enough to understand these relationships, and the balance of power in the sector, is essential. Likewise, it is important to have several contacts from the beneficiary country who understand the culture and history and can provide guidance in that context to ensure the right questions are asked.

Some of the right questions include:

- What do the beneficiaries want? Do they agree on the stated problem?
- Has the problem been addressed before?
- Did the previous solution(s) work? Why or why not?
- Why does the problem even need to be addressed?

The answers from the stakeholders should be heard and used to refine the design of the project.

Getting input from stakeholders can be done in various ways including surveys, facilitated workshops, and focus groups. These methods need to be implemented carefully with cultural context in mind. For instance, if a focus group is being conducted on the impact of an issue on women, it may be best to restrict the group to women-only so the women can speak freely.

It cannot be overemphasized how important political acceptance and involvement are to the success of development projects. While it may be a slow process to get political stakeholders on board, it is a necessary step and planning should not continue if this has not been accomplished.

7.2 Collect Data

Once the problem has been identified, the extent of the problem needs to be determined and this can be done through collection and analysis of data. In development projects, there is frequently insufficient data. However, the experience of this author has been that the data often does exist but it has to be patiently sought out and then converted to digital form. Rushed projects end up using data without much vetting. Wrong data is more dangerous than no data as it could point to nonexistent problems and, therefore, unnecessary solutions.

7.3 Refine Problem Statement

After getting input from relevant stakeholders, a specific goal should be set, keeping in mind the cultural context and the needs of the community. The goal should be achievable and measurable so that progress toward meeting it can be monitored and evaluated. Essentially, the goal should be clearly articulated so that it is clear what success means.

The boundaries of the problem should also be clearly defined to prevent scope creep and keep the solutions focused. The problem statement should be in plain English with no technical jargon in order to make it easier for experts from different fields to understand the problem and offer potential solutions.

7.4 Identify Solution Options

Once the problem is clearly defined, potential solutions can be identified. Solutions could come from technical experts, reading the literature, and/or anecdotal evidence from real world. In most cases, there will be several potential options. All implementable solutions should be documented. There is an important balance between not reinventing the wheel and not simply copying and pasting solutions. Innovation is required to contextualize solutions appropriately.

A maintenance plan should be included as part of the solution because maintenance is critical for the sustainability of a project. As such, brand-new technology is typically not appropriate for development work. Established technology is usually much less risky for several reasons. It does not turn the beneficiaries into guinea pigs. With established technology, maintenance is much easier since there will be local businesses that already understand the technology and can repair and upgrade the equipment as needed.

7.5 Compare Options

This step involves comparing the identified potential solutions side by side across the same metrics, weighing the advantages and disadvantages in light of the local context, and how closely they actually solve the defined problem.

A cost–benefit analysis should be carried out to determine which option offers the greatest benefits. A cost-effectiveness analysis should be carried out to determine which option has the greatest benefits for the expenditure or which option has the lowest cost for the expected outcome. Simplicity is key. The best solutions are not necessarily the most complicated.

Some of the metrics to be considered include the following:

- The context, e.g., rural vs. urban setting, environment, policies, culture, and history
- The downsides and advantages of each option
  - Cost – capital, operating, and maintenance costs. Normalized costs such as costs per kWh may be useful in addition to absolute costs
  - Access by beneficiaries especially protected groups like women and girls
  - Timelines – how long the solution will take to plan and implement
  - Appropriate scale – for instance, a nuclear power plant may not be appropriate for a village of 100 people
  - Supply chain – how easy it is to get materials to the area for establishing the solution
  - Convenience – how easy it is to get materials to the area for continued operations
  - How much maintenance is required to keep the solution working
  - Whether technology is established one
- The effects on the environment and the cost of mitigating them
- The impacts on the local community
  - The cost and timeline for resettlement
  - The cost of buying land, if necessary
  - Effects on local enterprises
  - Impacts on women and other minority, marginalized, or vulnerable groups

Once again, the input of the relevant stakeholders should be sought. Aside from technological considerations, the stakeholders may be able to provide cultural and historical context that will favor some solutions over others in terms of sustainability.

7.6 Select Option

After input from the experts and stakeholders, a decision must be made on one solution that will be carried forward. It is useful to rank the solutions using the metrics from the previous step, in case, for any reason, the number one solution cannot be implemented. The number two solution will be the next obvious choice.

7.7 Develop Plan

A plan most now be developed for the selected option. The plan should include contingencies for identified risks. Metrics to be used for M&E should be clearly identified and plans for measuring these metrics should be established. Additionally, it is important to model different technical or economic scenarios of the chosen solution and to make the solution as modular as possible so some parts can be dropped if resources run out. Factors that can be varied for the models include fuel costs, length of transmission line, and time ranges for forecasted factors like energy demand.

Sufficient resources, in terms of money and labor, must be allocated for the execution of the project. Subject-matter experts should be identified to drive the development and implementation of the project. Future maintenance should be a part of the plan to ensure the sustainability of the project. Unintended consequences must be mitigated by having locals heavily involved in the planning. Resilience should be built into the plans. This could mean building in redundancies to ensure, for example, that the transmission line or power plant can withstand severe storms that only hit rarely.

It is important to ensure that local firms can participate in the planning and implementation of the project to promote capacity development. Most development work is done by international firms that are largely based in the western world. Local firms are pushed out because they typically do not have experience at the scales required for development projects in terms of money spent and years of experience. It would be helpful if requirements were made for these international firms to partner with local firms to give them experience that they can add to their portfolio to allow them to compete for future development jobs. This practice would lead to local capacity development which would ensure sustainability of the projects in the country.

7.8 Execute Plan

It is important to include locals on planning and execution teams to help navigate the unwritten rules for getting things done in the country. Continuous check-ins with local stakeholders are also important for metrics tracking and to maintain a sense of ownership in the local stakeholders.

As unexpected factors are encountered, the contingency plans developed in the previous step can be implemented. The M&E team should keep track of the relevant metrics and keep relating them back to the original problem to ensure the root issue is being addressed.

7.9 Closing

M&E of the project needs to continue even after execution is complete and should ideally be done by an independent institution. Metrics agreed upon in the planning stages should be measured over time to determine whether or not the project is a success, and, if so, to what degree. Lessons learned even from failed projects can be applied to make future projects more sustainable. Input from the beneficiaries should be taken into consideration. The project and lessons learned should be documented and archived.

Every project is different, so the steps in this framework may need to be modified depending on the scale of the project, the type of development institution, infrastructure vs. policy reform, and so on. It is also important to note that this is not a project development framework, but a framework for sustainable development. More detailed project development frameworks can be made to fit into this framework to ensure sustainability.

Chapter 8
Project Examples

In this chapter, we examine examples of energy development projects that are unsustainable and sustainable. There are always unintended consequences with every project. Unsustainability here will mostly refer to the project not achieving its intended goal or the lack of sustainment of achieved goals. As explained in Chapter 7, the beneficiaries are usually not included in most of the planning. The beneficiaries are more likely to accept, protect, and maintain the solution if they feel like they have some ownership in it [84].

8.1 Unsustainable Energy Projects

8.1.1 Low-Cost Connections in Tanzania

Country Problem
From 2008 to 2013, the electricity connection rate in Tanzania was only about 18% nationally, and under 4% in the rural areas [85].

Proposed Solution
The Millennium Challenge Corporation (MCC) funded grid extensions in Tanzania from 2008 to 2013. Connection fees were thought to be a barrier for connecting to the grid, so MCC funded low-cost connection offers.

Project Details
MCC's low-cost connection offers reduced connection fees by at least 80% in 27 out of 178 communities getting new grid lines [85].

Results
Low-cost connection offers are estimated to have increased connections by 13%. It was assumed that within 1 year of the grid being extended, 35,000 new connections would be realized. Despite the availability of low connection costs, only 10,794 connections were realized up to 3 years after grid construction. This low-connection rate suggests that connection costs were not the main issue preventing people from connecting to the grid when it was available.

https://doi.org/10.1515/9783111643489-008

Lessons

Studies need to be carried out to understand consumer behavior and what incentives are needed to get people to connect to the grid. Outreach efforts could be made to market the project and raise awareness of the project and any accompanying incentives.

8.1.2 High Maintenance Cost for Solar Panels in a Burkina Faso Village

Country Problem

There was no electricity in a village in Burkina Faso called Bazoulé, roughly 23 km southwest of Ouagadougou, the capital.

Proposed Solution

This author observed that a European NGO had funded solar panel installations on the roofs of some huts in the village.

Project Details

The single solar panel installations were used to power a single lightbulb which was generally used by school children to do their homework. Maintenance of the solar system was required every 3 months.

Results

Unfortunately, the cost of maintenance was about three times the average monthly earnings of the villagers. As a result, villagers had to save up for maintenance and could generally only afford to have the equipment serviced once a year. This will contribute to shortening the life of the equipment.

Lessons

It is questionable whether this huge investment into maintenance is worthwhile for lighting a single lightbulb. Project planning should always include maintenance after installation is complete and the associated costs. A solar panel farm where the villagers pool their money together and share costs may be more sustainable and less of a burden on each individual family.

8.1.3 Expensive Electricity in Lesotho

Country Problem

In Lesotho in 1994, only about 0.01% of the population had access to electricity.

Proposed Solution

In 1986, the World Bank, the European Investment Bank, and the African Development Bank funded the $3.5 billion Lesotho Highlands water project to supply hydroelectric power. Construction on the power station ran from 1994 to 1998.

Project Details

The project diverted fresh water from the mountains using a series of dams to provide freshwater and generate electricity. The Muela power station, fed from the Katse Dam on the Malibamat'so River, was designed to supply 72 MW of hydroelectric power to Lesotho.

Results

The cost of electricity generated from the project was too high for the consumers in Lesotho at $12.29 per month compared to $2 a month normally spent on fuel like wood [86]. Additionally, the diversion of water caused issues for the environment and negatively affected the economies of communities downstream of the project. Nonetheless, electricity access in 2019 had increased to about 49% [87].

The fallout from the project included the conviction of three of the world's largest construction firms on corruption charges and the jailing of the project's CEO. Compensation has only been provided in kind and to only a few hundred households impacted by the dams. The dams took away farms, fields, and trees that the local population depended on to make a living. Conflicts have arisen due to the scarcity of grazing land. Springs have dried up due to construction activity, forcing villagers to travel greater distances to collect water. Accessing water from the dams is illegal as the dams are South African property. Compensation policies favored men and excluded households which did not have men present [86].

Lessons

One lesson learned from this example is that thorough environmental and social impact studies should be mandatory for any development project. Fair compensation should be built into the budget for impacts like loss of income or resettlement. Political will is essential for successful resettlement. This example also highlights the importance of engaging with the local community to ensure that negative impacts are minimized. Root cause analysis is needed to get to the root of a problem so that proposed solutions actually impact the identified problem.

8.2 Sustainable Energy Projects

8.2.1 Azito Power Plant Expansion in Cote d'Ivoire

Country Problem

The national electrification rate of Cote d'Ivoire in 2011 was 59%.

Proposed Solution

The 288 MW Azito Power Plant was first constructed in 1999 as a simple cycle gas-fired power station. In 2011, a concession agreement was made between Azito Energie and the Government of Cote d'Ivoire to upgrade the plant to a combined cycle and increase the installed capacity to 426 MW by addition of a 138 MW unit [88].

Project Details

Funding for the expansion project, estimated to cost approximately $450 m, came from the International Finance Corporation (IFC), and various European development finance institutions led by Proparco and the West African Development Bank (BOAD). Azito. Equity finance for the project was provided by Energie Holding and Globeleq.

The expansion was done under a Build–Own–Operate–Transfer (BOOT) scheme. Electricity would be sold to Compagnie Ivoirienne d'Electricité (CIE) through a 20-year concession. Expansion of the plant was completed in 2015.

Results

The overall efficiency of the plant increased from 29.5% to 44% with no increase in emissions or gas consumption. Roughly 1,300 jobs were created during expansion of the plant and 60% of them were filled from the local communities. The project added 138 MW of electricity generating capacity to the Ivorian grid. Azito power plant now accounts for 25% of Cote d'Ivoire's electricity capacity, which is an increase of 10%.

Lessons

The success of this project demonstrates that political will and policies that encourage private sector participation are essential for power sector projects to succeed.

8.2.2 The Rwamagana Solar Power Station in Rwanda

Country Problem

In 2013, Rwanda had a low electrification rate around 15.2%.

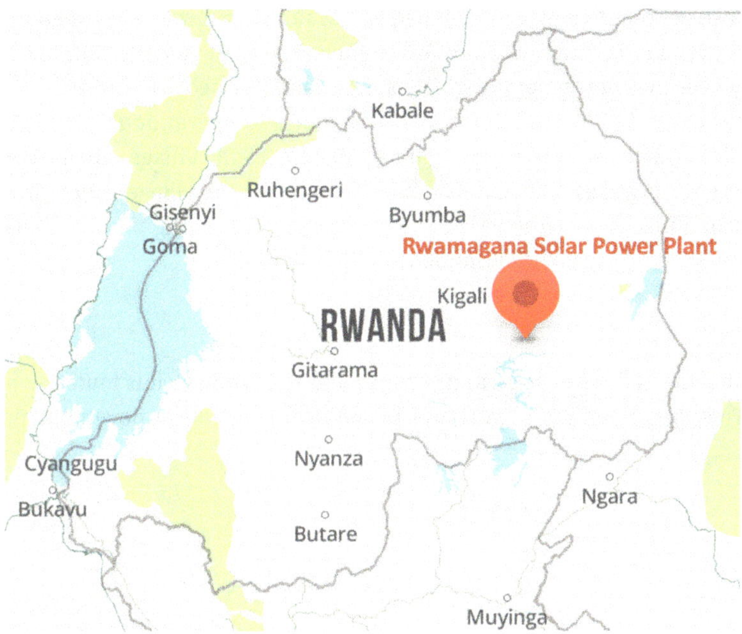

Figure 8.1: Location of Rwamagana Solar Power Plant [89].

Proposed Solution

GigaWatt Global (GWG) secured financing for a solar power station called Rwamagana, under a Build–Own–Operate model.

Project Details

Rwamagana is an 8.5-MW solar power station located 60 km east of Kigali as shown in Figure 8.1. It contains 28,360 solar panels on 50 acres in the shape of Africa [90]. It was funded in 2013 by the Netherlands Development Finance Company (FMO), the Emerging Africa Infrastructure Fund, Norfund, the United States Overseas Private Investment Corporation, and Finland's Energy and Environment Partnership.

The contract with GWG was signed in July 2013 and financial close was achieved in February 2014. The power station was connected to the grid in July 2014 and the station began operating at maximum capacity in September 2014.

Results

The station produced 15 GWh in its first year and powers more than 15,000 homes [91]. GWG will operate the power station for 25 years, after which ownership will transfer to the Government of Rwanda. The electricity generated is sold to the national electric utility Rwanda Energy.

The final construction price, which included access to the grid, was $23.7 million, just $0.7 million over the contract price of $23 million. The final construction price was only slightly over the contract price, which is unusual for projects of this scale.

The project created 280 jobs during construction and 15 operational jobs. The land for the project was leased from the Agahozo-Shalom Youth Village which was established to care for orphans. The lease supplies funding for the village which provides a home, schooling, and extracurricular activities for 512 young people. GWG also provides training on solar power to students in the village [92].

Lessons

The speed at which the deal was closed and construction was completed is touted as a great success. This illustrates the importance of political goodwill in development projects.

Glossary

Base load: the minimum amount of electricity demand over a specified period.

Build–Operate–Transfer (BOT): private sector participation mechanism in which a project is financed by the government but built and operated by the private sector, and ownership is transferred to the government at the end of the concession period.

Build–Own–Operate (BOO): ownership remains permanently with the private sector.

Build–Own–Operate–Transfer (BOOT): the contractor finances, builds, and operates a discrete asset for a set period of time.

Capacity factor: the ratio of actual electricity production of a power plant over a set period of time to its maximum possible electricity production over the same period.

Carbon credit: a certificate that gives the holder the right to emit a tonne of carbon dioxide or the equivalent amount of another greenhouse gas.

Concession: private sector participation mechanism in which the concessionaire is responsible for all operation, maintenance, and upgrading of infrastructure for a set period.

Demand management: voluntary and compensated program implemented to reduce consumer usage of electricity during high-system demand periods.

Dispatchability: when power output of a source of electricity can be adjusted to respond to electric grid demand.

Divestiture: permanent change of ownership of an asset to the private sector.

Duck curve: plot of the difference between electricity demand and solar power generation over the course of a day.

Electrification rate: percentage of the population with access to electricity. Also called electricity access.

Energy efficiency: the avoidance or reduction of energy waste.

Energy policy: comprehensive approach of an organization or country to production and supply of energy.

Energy sustainability: addressing today's needs without compromising the ability of society to address future energy needs.

Energy–water nexus: the interconnection between water and the production of energy.

Enhanced oil recovery: increasing the amount of oil recovered from a reservoir by injecting a liquid or gas into the reservoir.

Feed-in tariff (FiT): a fixed price paid to renewable energy producers for a set period of time for injecting their electricity into the grid.

Flexibility: the ability of a power system to respond to changes in electricity demand and supply.

Flue gas: exhaust gas produced by combustion of fuel at a power plant.

Grid cybersecurity: the art of protecting the electric grid from unauthorized access.

Grid integration of renewable energy: power system planning to maintain stability and reliability of the grid when integrating intermittent renewable energy.

https://doi.org/10.1515/9783111643489-009

Grid parity: when the LCOE of an alternative energy source is less than or equal to the cost the consumer would pay a utility for electricity.

Independent power producer (IPP): a nongovernmental entity which generates electricity for sale to utilities and other consumers.

Law: passed by the legislative body of the country and remains in place even when a new government takes power.

Lease contract: privatization mechanism in which the private sector operates and maintains the asset but is not responsible for financing.

Levelized cost of electricity (LCOE): the average net present cost of electricity generation over the lifetime of a generation plant

Microgrid: a small version of the grid that can operate independently of the main grid.

Net metering: a way in which a utility compensates renewable energy producers for excess electricity by providing credit for the amount of electricity sent back into the grid.

Nonrenewable energy: energy that cannot be naturally replenished within the lifetime of a human being.

Oxycombustion: combustion of fuel in pure oxygen rather than air.

Peak load: a spike in demand over a short period.

Policy: outlines the goals of the government and how they will be achieved.

Post-combustion capture: CO_2 capture from flue gas after combustion of fuel

Power pool: a mechanism for utilities to exchange electricity.

Power Purchase Agreement (PPA): a contract that allows one party to purchase electricity from another party which generates electricity.

Pre-combustion CO_2 capture: capture of CO_2 from combustion of syngas derived from coal.

Private sector participation (PSP): the engagement of the private sector with government in providing services that are expected to be provided by government.

Privatization: transfer of ownership or licensing of assets from government to the private sector.

Productive use of electricity (PUE): the use of electricity to produce outcomes that can be measured in monetary terms

Public–Private Partnership (PPP): government and the private sector working together to provide services to the public.

Regulations: rules that provide detailed direction to an implementing agency or other actor on how to implement laws.

Renewable energy: energy that can be replenished within the lifetime of a human being.

Service contract: mechanism by which a utility sources goods or services from the private sector

Smart grid: an electric grid that supports the flow of operational information through a communications network.

Smart meter: a meter that collects and transmits information related to electricity such as level of consumption, current, and voltage.

Stakeholder: a person or group affected by the outcome of a project.

Storage: capture of electricity generated at one time and kept for use at a later time.

Syngas: a fuel gas primarily comprised of carbon monoxide and hydrogen. Also called synthesis gas.

Utility: a company that is involved in generation, transmission, and/or distribution of electricity.

Vertical unbundling: separating a utility into separate companies that each focus on generation, transmission, or distribution of electricity.

References

[1] United Nations, "Affordable and Clean Energy," [Online]. Available: https://unstats.un.org/sdgs/report/2023/goal-07/. [Accessed 27 April 2025].

[2] P. Benoit, "Energy and Development in a Changing World: A Framework for the 21st Century," 4 March 2019. [Online]. Available: https://energypolicy.columbia.edu/research/energy-and-development-changing-world-framework-21st-century.

[3] P. J. Lloyd, "The role of energy in development," *Journal of Energy in Southern Africa*, vol. 28, no. 1, pp. 54–62, 2017.

[4] P. R. a. M. R. H. Ritchie, "Energy Production and Consumption," [Online]. Available: https://ourworldindata.org/energy-production-consumption. [Accessed 27 April 2025].

[5] World Bank, "Population, total – North America," [Online]. Available: https://data.worldbank.org/indicator/SP.POP.TOTL?end=2023&locations=XU&start=1960&type=shaded&view=chart. [Accessed 26 May 2025].

[6] World Bank, "Population growth (annual %) – Sub-Saharan Africa," [Online]. Available: https://data.worldbank.org/indicator/SP.POP.GROW?end=2023&locations=ZG&start=1961&view=chart. [Accessed 26 May 2025].

[7] H. R. a. M. Roser, "Access to Energy," Our World in Data, September 2019. [Online]. Available: https://ourworldindata.org/energy-access. [Accessed 3 June 2020].

[8] International Energy Agency, "World energy balances and statistics," [Online]. Available: https://www.iea.org/world/energy-mix. [Accessed 31 May 2025].

[9] The World Bank, "Access to electricity (% of population)," [Online]. Available: https://data.worldbank.org/indicator/EG.ELC.ACCS.ZS. [Accessed 10 May 2020].

[10] J. M. Carmody, "Rural Electrification in the United States," *The ANNALS of the American Academy of Political and Social Science*, vol. 201, no. 1, pp. 82–88, 1939.

[11] G. R. G. C. H. T. Dinh, "Performance of Manufacturing Firms in Africa An Empirical Analysis," The World Bank, Washington, DC, 2012.

[12] H. T. D. a. G. R. G. Clarke, "Performance of Manufacturing Firms in Africa," 2012. [Online]. Available: http://documentos.bancomundial.org/curated/es/410911468211167052/pdf/717320PUB0Publ067869B09780821396322.pdf. [Accessed 6 May 2020].

[13] International Energy Agency, "Africa: Sources of electricity generation," [Online]. Available: https://www.iea.org/regions/africa/electricity. [Accessed 21 June 2025].

[14] The World Bank, "Electric power transmission and distribution losses (% of output)," [Online]. Available: https://data.worldbank.org/indicator/EG.ELC.LOSS.ZS. [Accessed 21 June 2025].

[15] Clean Energy Institute, "Lithium-Ion Battery: What is a lithium-ion battery and how does it work?," University of Washington, [Online]. Available: https://www.cei.washington.edu/education/science-of-solar/battery-technology/. [Accessed 7 June 2020].

[16] S. Silbir, "Energy in natural processes and human consumption – some numbers," 2005. [Online]. Available: https://www.google.com/url?sa=t&source=web&rct=j&opi=89978449&url=https://www.researchgate.net/profile/Selim-Silbir/post/What-is-calorific-value-of-ethanol-in-MJ-l/attachment/5adb02134cde260d15da14e4/AS%253A617783008178178%25401524302355633/download/energ. [Accessed 21 June 2025].

[17] World Bank Group, "Electricity production from coal sources (% of total) – Sub-Saharan Africa," [Online]. Available: https://data.worldbank.org/indicator/EG.ELC.COAL.ZS?locations=ZG. [Accessed 21 June 2025].

[18] US Energy Information Agency, "Coal explained," [Online]. Available: https://www.eia.gov/energyexplained/coal/. [Accessed 1 June 2020].

[19] Environmental Protection Agency, "Acid Rain Program Results," [Online]. Available: https://www.epa.gov/acidrain/acid-rain-program-results. [Accessed 7 June 2020].

https://doi.org/10.1515/9783111643489-010

[20] Energy Information Agency, "Electricity explained. How electricity is generated," [Online]. Available: https://www.eia.gov/energyexplained/electricity/how-electricity-is-generated.php. [Accessed 11 June 2020].

[21] National Energy Technology Lab, "Water Vulnerabilities for Existing Coal-fired Power Plants," August 2010. [Online]. Available: https://publications.anl.gov/anlpubs/2010/08/67687.pdf. [Accessed 23 May 2020].

[22] World Bank, "Electricity production from oil sources (% of total)," World Bank, [Online]. Available: https://data.worldbank.org/indicator/EG.ELC.PETR.ZS?view=map. [Accessed May 31 2020].

[23] Department of Ecology, "Hanford Quick Facts," State of Washington, [Online]. Available: https://web.archive.org/web/20080624232748/http://www.ecy.wa.gov/features/hanford/hanford facts.html. [Accessed 24 June 2020].

[24] Scientific American, "Clearing the Radioactive Rubble Heap That Was Fukushima Daiichi, 7 Years On," 9 March 2018. [Online]. Available: https://www.scientificamerican.com/article/clearing-the-radioactive-rubble-heap-that-was-fukushima-daiichi-7-years-on/. [Accessed 29 June 2025].

[25] "Plant Vogtle: The True Cost of Nuclear Power in the United States," [Online]. Available: https://gcvedfund.org/wp-content/uploads/2025/03/Vogtle-Truth-Report_update_final-1.pdf. [Accessed 29 June 2025].

[26] World Nuclear Association, "Nuclear Power in South Africa," [Online]. Available: https://www.world-nuclear.org/information-library/country-profiles/countries-o-s/south-africa.aspx. [Accessed 24 June 2020].

[27] World Bank, "Electricity production from nuclear sources (% of total) – Sub-Saharan Africa," [Online]. Available: https://data.worldbank.org/indicator/EG.ELC.NUCL.ZS?locations=ZG&view=map. [Accessed 29 June 2025].

[28] World Bank, "Electricity production from hydroelectric sources (% of total)," [Online]. Available: https://data.worldbank.org/indicator/EG.ELC.HYRO.ZS?view=map. [Accessed 31 May 2020].

[29] ESI Africa, "Hydropower at risk as water levels at Lake Victoria rise," 4 May 2020. [Online]. Available: https://www.esi-africa.com/industry-sectors/generation/hydropower-at-risk-as-water-levels-at-lake-victoria-rise/. [Accessed 18 June 2020].

[30] H. M. a. S. J. Zarrouk, "Efficiency of Geothermal Power Plant: A Worldwide Review," November 2012. [Online]. Available: https://www.geothermal-energy.org/pdf/IGAstandard/NZGW/2012/46654final00097.pdf. [Accessed 31 May 2020].

[31] Boston University – Institute for Global Sustainability, "Power plant efficiency since 1900," 24 July 2023. [Online]. Available: https://visualizingenergy.org/power-plant-efficiency-since-1900/. [Accessed 3 August 2025].

[32] E. Y. K. a. J. Muguthu, "Geothermal Energy Development in East Africa: Barriers and Strategies," *Journal of Energy Research and Reviews*, vol. 2, no. 1, pp. 1–6, 2019.

[33] KenGen, "KenGen – Home," [Online]. Available: https://kengen.co.ke. [Accessed 3 August 2025].

[34] M. Zastrow, "South Korea accepts geothermal plant probably caused destructive quake," Nature, 22 March 2019. [Online]. Available: https://www.nature.com/articles/d41586-019-00959-4. [Accessed 31 May 2020].

[35] ESMAP, "Global Solar Atlas," World Bank, [Online]. Available: https://globalsolaratlas.info. [Accessed 10 June 2020].

[36] National Renewable Energy Lab NREL, "Solar Installed System Cost Analysis," [Online]. Available: https://www.nrel.gov/solar/market-research-analysis/solar-installed-system-cost. [Accessed 3 August 2025].

[37] D. F. a. R. M. R. Fu, "U.S. Solar Photovoltaic System Cost Benchmark: Q1 2018," National Renewable Energy Lab, Golden, CO, 2018.

[38] Dubai Electricity and Water Authority, "Mohammed bin Rashid Al Maktoum Solar Park," [Online]. Available: https://www.dewa.gov.ae/en/about-us/strategic-initiatives/mbr-solar-park. [Accessed 28 September 2025].

[39] ESMAP, "Global Wind Atlas," World Bank, [Online]. Available: https://globalwindatlas.info. [Accessed 10 June 2020].

[40] Scientific American, "Bat Killings by Wind Energy Turbines Continue," 2016. [Online]. Available: https://www.scientificamerican.com/article/bat-killings-by-wind-energy-turbines-continue/. [Accessed 30 May 2020].

[41] Energy Information Agency, "Average U.S. electricity customer interruptions totaled nearly 8 h in 2017," [Online]. Available: https://www.eia.gov/todayinenergy/detail.php?id=37652. [Accessed 18 June 2020].

[42] C. S. L. a. M. D. McCulloch, "Levelized Cost of Energy for PV and Grid Scale Energy Storage Systems," 2016. [Online]. Available: https://arxiv.org/abs/1609.06000. [Accessed 9 May 2020].

[43] Energy Information Agency, "Levelized Cost and Levelized Avoided Cost of New Generation Resources in the Annual Energy Outlook 2020," February 2020. [Online]. Available: https://www.eia.gov/outlooks/aeo/pdf/electricity_generation.pdf. [Accessed 9 May 2020].

[44] CAISO, "Today's Outlook," 18 June 2020. [Online]. Available: http://www.caiso.com/TodaysOutlook/Pages/default.aspx. [Accessed 18 June 2020].

[45] E. Martinot, "Grid Integration of Renewable Energy: Flexibility, Innovation, Experience," *Annual Review of Environment and Resources*, 2016.

[46] Statista, "World carbon dioxide emissions from 2008 to 2018, by region (in million metric tons of carbon dioxide)," [Online]. Available: https://www.statista.com/statistics/205966/world-carbon-dioxide-emissions-by-region/. [Accessed 7 June 2020].

[47] R. S. Pindyck, "Climate Change Policy: What Do the Models Tell Us?," *Journal of Economic Literature*, vol. 51, no. 3, pp. 860–872, 2013.

[48] P. J. Meier, "Life-Cycle Assessment of Electricity Generation Systems and Applications for Climate Change Policy Analysis," University of Wisconsin-Madison, Madison, WI, 2002.

[49] L. E. Ø. a. S. H. P. Kvam, "Comparison of energy consumption for different CO2 absorption configurations using different simulation tools," *Energy Procedia*, vol. 63, pp. 1186–1195, 2014.

[50] J. E. D. H. J. H. E. S. Rubin, "The cost of CO2 capture and storage," *International Journal of Greenhouse Gas Control*, 2015.

[51] Food and Agricultural Organization, "AQUASTAT – FAO's Global Information System on Water and Agriculture," [Online]. Available: http://www.fao.org/aquastat/en/overview/methodology/water-use. [Accessed 2 June 2020].

[52] WorldOMeter, "Global Water Use," 2014. [Online]. Available: https://www.worldometers.info/water/. [Accessed 7 June 2020].

[53] M. K. A. M. A. H. W. Z. W. I. C. M. Chew, "Practical performance analysis of an industrial-scale ultrafiltration membrane water treatment plant," *Journal of The Taiwan Institute of Chemical Engineers*, vol. 46, pp. 132–139, 2015.

[54] Emerging Technologies, "Ultraviolet Disinfection of Water and Wastewater," [Online]. Available: http://e3tnw.org/ItemDetail.aspx?id=13. [Accessed 6 May 2018].

[55] EMIS, "Nanofiltration," [Online]. Available: https://emis.vito.be/en/bat/tools-overview/sheets/nanofiltration. [Accessed May 2018].

[56] LennTech, "Disinfection by Ultraviolet Light," [Online]. Available: https://www.lenntech.com/library/uv/will1.htm. [Accessed 5 May 2018].

[57] Water Conditioning & Purification Magazine, "High-Performance Nitrate Removal Ion Exchange System," 21 November 2010. [Online]. Available: http://wcponline.com/2010/11/21/high-performance-nitrate-removal-ion-exchange-system/. [Accessed 5 May 2018].

[58] S. W. S. H.-W. M. Y. J. L. a. P.-C. C. S.-Y. Pan, "Development of a Resin Wafer Electrodeionization Process for Impaired Water Desalination with High Energy Efficiency and Productivity," *ACS Sustainable Chemistry & Engineering*, vol. 5, no. 4, pp. 2942–2948, 2017.

[59] R. N. G. H. a. K. C. H. J. Macknick, "Operational water consumption and withdrawal factors for electricity generating technologies: a review of existing literature," *Environmental Research Letters*, vol. 7, no. 4, 2012.

[60] K. T. S. R. A. M. Peer, "Characterizing cooling water source and usage patterns across US thermoelectric power plants: a comprehensive assessment of self-reported cooling water data," *Environmental Research Letters*, vol. 11, 2016.

[61] K. J F. A. H.-L. A L. J. M. N. M J. R a. S. T. Averyt, "Freshwater use by U.S. power plants: Electricity's thirst for a precious resource," Union of Concerned Scientists, Cambridge, MA, 2011.

[62] D. W. E. K. B. Bernard, "Carbon Markets: Have They Worked for Africa?," *Review of Integrative Business and Economics Research*, vol. 6, no. 2, pp. 90–104, 2017.

[63] J. Monteagudo, "Power Grid Cybersecurity – where are we now?," Cybersecurity Observatory, [Online]. Available: https://cyberstartupobservatory.com/power-grid-cybersecurity-where-are-we-now/. [Accessed 12 June 2020].

[64] Bipartisan Policy Center's Electric Grid Cybersecurity Initiative, "Cybersecurity and the North American Electric Grid: New Policy Approaches to Address an Evolving Threat," February 2014. [Online]. Available: https://bipartisanpolicy.org/wp-content/uploads/2019/03/Cybersecurity-Electric-Grid-BPC.pdf. [Accessed 7 June 2020].

[65] West African Power Pool, "Creation of the WAPP," [Online]. Available: http://www.ecowapp.org/en/countries. [Accessed 6 June 2020].

[66] Southern African Power Pool, "Demand and Supply," [Online]. Available: http://www.sapp.co.zw/demand-and-supply. [Accessed 6 June 2020].

[67] Central African Power Pool, "Members," [Online]. Available: https://www.peac-sig.org/en/members.html. [Accessed 6 June 2020].

[68] Eastern Africa Power Pool, "The Eastern Africa Power Pool," [Online]. Available: http://eappool.org. [Accessed 6 June 2020].

[69] Southern African Power Pool, "SAPP Existing Generation Stations – 2015/16," [Online]. Available: http://www.sapp.co.zw/sites/default/files/R9%20-%20SAPP%20Statistics%20-%202016.pdf. [Accessed 6 June 2020].

[70] West African Power Pool, "Key Indicators," [Online]. Available: http://icc.ecowapp.org/list-stat. [Accessed 6 June 2020].

[71] Africa Energy Portal, "Power pools enabling SSA's transmission corridors," 16 May 2019. [Online]. Available: https://africa-energy-portal.org/blogs/power-pools-enabling-ssas-transmission-corridors. [Accessed 23 June 2020].

[72] R. White, "Workshop on Productive Uses of Renewable Energy – Synthesis and Report," GEF-FAO, Washington, DC, 2003.

[73] World Bank, "Rural Electrification: a hard look at costs and benefits. Operations Evaluation Department, Precis Number: 90," World Bank, Washington, DC, 1995.

[74] "Why we should pull the plug on privatising electricity," The Conversation, 25 November 2012. [Online]. Available: https://theconversation.com/why-we-should-pull-the-plug-on-privatising-electricity-10824. [Accessed 7 June 2020].

[75] A. F. A. A. L. ROCHA, "Private versus public electricity distribution utilities: Are outcomes different for end-users?," World Bank, 3 May 2018. [Online]. Available: https://blogs.worldbank.org/developmenttalk/private-versus-public-electricity-distribution-utilities-are-outcomes-different-end-users. [Accessed 6 June 2020].

[76] World Bank Group, “PPP Arrangements / Types of Public-Private Partnership Agreements,” World Bank, 12 September 2019. [Online]. Available: https://ppp.worldbank.org/public-private-partnership/agreements. [Accessed 7 June 2020].

[77] P. G. a. M. Kerf, “Concessions – The Way to Privatize Infrastructure Sector Monopolies,” *Public Policy for the Private Sector – The World Bank*, October 1995.

[78] World Bank Group, “Concessions, Build-Operate-Transfer (BOT) and Design-Build-Operate (DBO) Projects,” PUBLIC-PRIVATE-PARTNERSHIP LEGAL RESOURCE CENTER, 8 February 2018. [Online]. Available: https://ppp.worldbank.org/public-private-partnership/agreements/concessions-bots-dbos. [Accessed 7 June 2020].

[79] ESMAP, “Private Sector Participation in Electricity Transmission and Distribution,” World Bank, Washington, DC, 2015.

[80] I. P.-A. R. J. S. D. N. G. Jacquot, “Assessing the potential of electrification concessions for universal energy access: Towards integrated distribution frameworks,” Massachusetts Institute of Technology Energy Initiative, Cambridge, MA, 2019.

[81] H. Slim, “Is racism part of our reluctance to localise humanitarian action?,” Humanitarian Practice Network, 5 June 2020. [Online]. Available: https://odihpn.org/blog/is-racism-part-of-our-reluctance-to-localise-humanitarian-action/. [Accessed 7 June 2020].

[82] L. Cornish, “Q&A: Degan Ali on the systemic racism impacting humanitarian responses,” Devex, 20 June 2019. [Online]. Available: https://www.devex.com/news/q-a-degan-ali-on-the-systemic-racism-impacting-humanitarian-responses-95083. [Accessed 23 June 2020].

[83] B. Omakwu, “Opinion: On equity in the international development sector – we need more intravists,” Devex, 5 June 2020. [Online]. Available: https://www.devex.com/news/opinion-on-equity-in-the-international-development-sector-we-need-more-intravists-97404. [Accessed 20 June 2020].

[84] P. C. S. J. V. H. P. B. M. E. C. X. Ikejemba, “Failures & generic recommendations towards the sustainable management of renewable energy projects in Sub-Saharan Africa (Part 2 of 2),” *Renewable Energy*, vol. 113, pp. 639–647, 2017.

[85] A. M. A. P. J. S. D. V. K. B. H. B. L. M. A. D. C. K. a. T. C. D. Chaplin, “Grid Electricity Expansion in Tanzania by MCC: Findings from a Rigorous Impact Evaluation,” Mathematica Policy Research, Washington, DC, 2017.

[86] R. Hoover, “Pipe Dreams The World Bank's Failed Efforts to Restore Lives and Livelihoods of Dam-Affected People in Lesotho,” International Rivers Network, Berkeley, CA, 2001.

[87] EnergyPedia, “Lesotho Energy Situation,” [Online]. Available: https://energypedia.info/wiki/Lesotho_Energy_Situation. [Accessed 23 June 2020].

[88] Power Technology, “Azito Power Plant Expansion, Abidjan,” Power Technology, [Online]. Available: https://www.power-technology.com/projects/azito-power-plant-expansion-abidjan/. [Accessed 7 June 2020].

[89] Waze, [Online]. Available: https://www.waze.com/livemap/directions?latlng=−2.026111%2C30.377222&utm_expid=.K6QI8s_pTz6FfRdYRPpI3A.0&utm_referrer=&zoom=15. [Accessed 15 June 2020].

[90] D. Smith, “How Africa's fastest solar power project is lighting up Rwanda,” The Guardian, 23 November 2015. [Online]. Available: https://www.theguardian.com/environment/2015/nov/23/how-africas-fastest-solar-power-project-is-lighting-up-rwanda. [Accessed 5 June 2020].

[91] FMO Entrepreneurial Development Bank, “A Solar Farm for Rwanda,” [Online]. Available: http://static1.squarespace.com/static/55a36893e4b0e8589c3430bb/t/55c46687e4b0c8f840711190/1438934663423/Client+Case+Gigawatt+Global+Short.pdf. [Accessed 7 June 2020].

[92] GigaWatt Global, "Rwanda gets a solar field," [Online]. Available: https://gigawattglobal.com/projects3/rwanda/. [Accessed 7 June 2020].

[93] A. A. a. A. Arar, "Performance Assessment and Improvement of Central Receivers Used for Solar Thermal Plants," *Energies*, vol. 12, no. 3079, 2019.

[94] UNEP RISO Centre, "Carbon Markets and Africa: A Quick Fact Sheet for Journalists," 2012. [Online]. Available: https://www.afdb.org/fileadmin/uploads/afdb/Documents/Generic-Documents/Carbon%20Market%20Quick%20Facts%20%20ACF%202012.pdf. [Accessed 23 June 2020].

List of Figures

https://doi.org/10.1515/9783111643489-011

Index

https://doi.org/10.1515/9783111643489-012

www.ingramcontent.com/pod-product-compliance
Lightning Source LLC
Chambersburg PA
CBHW081551220326
41598CB00036B/6633

* 9 7 8 3 1 1 1 6 4 3 2 7 4 *